Preface

Cooling equipment is used in nearly all fields of human endeavor. While the house air conditioner, refrigerator, and freezer are the forms most readily known to the homeowner, a multitude of other machines using similar principles is essential for the operation of commerce and industry. For example, the modern food and fish processing businesses could not function at all without a means of chilling their product. Refrigeration is used in the tempering of alloys and the manufacture of drugs and a host of other products. Indeed, the applications of cooling equipment are so widespread that Western industrial civilization could not manage without them.

The basic principles are the same whatever the size of the machine may be. This text explains in understandable terms how an electric motor can produce a cooling effect and describes simple remedies for the more common malfunctions. It is apparent that with this information the homeowner can save himself considerable inconvenience, not to mention the loss of food by spoilage. In the matter of air conditioning, various types of equipment are described, and it is explained how an appropriate type of unit may be chosen. A fair amount of space is devoted to the related field of heat pumps, which are gaining wide acceptance in many parts of the country. Combination resistance heat and cooling units for mobile homes are discussed, with attention given to the special problems found in this type of housing.

This book also shows how some of the more complicated repairs may be made, to assist those who would like a more complete grasp of the subject. It explains the operation of the gas refrigerator, which uses heat to produce cold, and takes a look at interesting new developments, not the least of which is the exciting prospect of cooling homes with the energy supplied by the heat of the sun.

The author wishes to thank Ms. Pat Patrick and Mr. Don Allex of the Frigidaire Corporation, Dayton, Ohio, for their invaluable assistance in the preparation of this book.

BERNARD LAMERE

Guide to
Home Air Conditioners
and
Refrigeration Equipment

Revised Edition

Bernard Lamere

HAYDEN BOOK COMPANY, INC.
Rochelle Park, New Jersey

Library of Congress Cataloging in Publication Data

Lamere, Bernard.
  Guide to home air conditioners and refrigeration
equipment.

  Originally published: New York: J.F. Rider, 1963.
  Includes index.
  1. Dwellings—Air conditioning. 2. Refrigerators.
I. Title.
TH7688.H6L3 1981   621.5'6   81-6508
ISBN 0-8104-0956-9   AACR2

Printed in the United States of America

2	3	4	5	6	7	8	9	PRINTING
82	83	84	85	86	87	88	89	YEAR

Contents

Guide to
Home Air Conditioners
and
Refrigeration Equipment

Revised Edition

1

How Cooling Equipment Works

The value of ice for the preservation of food has been recognized for hundreds of years. In these days, when ice cubes can be made in the common household refrigerator, it is hard to imagine that, years ago, ice was so sought after that the exportation of natural ice from New England ports became a major item of commerce. Around 1810 Cuba was paying the top rate, $500 per ton!

Yankee ship owners hired hard-bitten sea captains who, in consideration of the fantastic profits, drove their vessels mercilessly to and from the ice fields. When it is realized that Boston, one of a number of ports in the business, shipped 65,000 tons of natural ice in one year, it is apparent that a large number of people were engaged in the cutting, handling, and transportation of this now-common commodity.

But, even as the New England ice shippers built their mansions, the seeds of their financial undoing had been sowed. The beginning of it was in the thinking of some European physicists in the years before the American Revolution.

They knew that when a kettle of water was placed over a fire, the heat would make the water boil and it would all eventually turn to vapor, disappearing in the air. And, they knew that if they held a cold piece of sheet metal over the rising vapor, drops of water would accumulate on the metal (see Fig. 1-1).

On the theory that all liquids act in the same way, they stated a law of physics: heat is required to make a liquid boil away (vaporize), and cooling will cause the vapor to return to its original liquid state (condense). They found this was generally true, though some liquids, such as ammonia and alcohol, would vaporize at an exceedingly low temperature. They gathered that if a kettle of alcohol would vaporize *without* being put over a fire, it must be *absorbing* heat from the surrounding air (see Fig. 1-2).

From other experiments they learned that the vapor occupies a considerably larger space than the original liquid. Or, to state the reverse of it: the volume of a vapor is reduced when it is cooled to a liquid. They further

1

Fig. 1-1   Water vapor (steam) will condense on a cold object

Fig. 1-2   Alcohol will vaporize from the heat of the surrounding air

theorized that if the vapor could be mechanically "squeezed," it would give off the heat that it must have absorbed.

They concluded that if they could take a liquid with a very low boiling (or vaporizing) point and put it in a container so fitted that the vapor would feed into a mechanical "squeezing device," pushing the somewhat compressed and much warmer vapor into a sort of radiator where it could give off its heat, the vapor should return to a liquid condition. Then, by arranging for the liquid to flow back into the vaporization chamber and repeat the cycle continuously, one could utilize the cold area of the vaporization chamber for cooling purposes.

They built a device using this theory, using ammonia (which will vaporize at a temperature below the freezing point of water) as the liquid or "refrigerant." It worked—the machine proved the theory to be correct. This principle of operation has been used in mechanical cooling equipment ever since.

To say that the first ice machines were not an unqualified success is an understatement. The ammonia had a corrosive effect on the cast iron parts, resulting in all sorts of malfunctions. After a while many bronze and brass parts were used because they would not "freeze" together the way iron did. By 1855, equipment had been devised that performed nicely, finding extensive use in breweries and meat-packing plants.

On this side of the ocean, a Mr. Twining was trying to build and sell ice machines. He did not have much success because, for some reason, people had the misconception that artificial ice was unhealthy. He did open a plant in Ohio, but he found general opposition to machine-made ice throughout the Northeast.

Before the Civil War, the Southerners had depended on natural ice from New England. With the advent of hostilities, the supply was shut off by the blockade. As the war progressed, more soldiers were dying of disease than bullets, and ice was desperately needed to treat fever sufferers, particularly those stricken with malaria.

It happened, as an accident of history, that one of the two countries that did not observe the Neutrality Law, made the ice machine available to the Southerners. The Rebels were faced with a difficult choice of whether to buy ice machines or gunpowder. They decided to purchase some ice machines. And so the firm of Carre, Ltee., in Paris, had some of its crates labeled: "For export to the Confederate States of America."

The invading Union Army found artificial ice being used in hospitals, and they learned that it did not make the patients worse than they were before. When word of this spread to the North, it helped to break down the widespread prejudice against machine-made ice.

In the decades after the war, there were occasional winters in New England that were warmer than usual, with the result that natural ice was in short supply the following summers. Doctors began to advise that it was safe to use artificial ice. By the 1890s, a considerable number of "ammonia plants" were in operation.

These ammonia machines were powered by steam because that was the most reliable source of power available at that time. Because a steam plant has to be constantly attended, the ice had to be produced on a large scale to be a financial success.

Refrigerators for the Home

It was apparent to the manufacturers of these machines that a small scale version of the ice machine for food-preservation purposes in the home would be very desirable. The development of reliable, self-starting, electric motors brought this dream closer to reality because the early manual-starting motors were of little value for home refrigeration.

The next step was to decide on the physical arrangement of the parts inside the cabinet. They knew the housewife did not want to freeze all her food; most of it merely needed chilling to 40° F. or so. Therefore, it followed that the freezing should take place in a relatively small box (for freezing water or ice cream), and the box could chill the air in the remaining part of the cabinet. And, because cold air settles, the "freeze box" had to be at the top.

The big, commercial ice machines used ammonia as the substance, or "refrigerant," that was pumped around through the system of tubing to produce a freezing temperature at the desired point. The ammonia worked very well, but it was apparent that it was not practical to use in a small household model. After the ammonia turns to vapor, it is very difficult to make it return to the liquid state. So difficult, in fact, that the equipment for mechanically "squeezing," or compressing, the vapor had to be so massive that, if used in a home, it would occupy the whole kitchen.

At that point, a technological "breakthrough" was achieved. Chemists had known for a long time that a gas called sulphur dioxide could be used in place of ammonia. It readily becomes a liquid when "squeezed," or compressed, and it gives off heat in the process. Then, as a liquid, it evaporates (vaporizes) to a gas, absorbing heat in the process. In other words, it functions as a refrigerant. They thought that, if "refrigerant" meant "a substance used to produce cold," the entire unit should be termed a "refrigerator." The first household refrigerators were marketed in the years after World War I, and, though they have changed in appearance and are vastly improved in efficiency, the basic principle of operation remains the same.

Cooling System Components

People proceeded to assign names to the refrigerator components. The box-shaped container in which the refrigerant vaporized (or evaporated) was, technically, an evaporator, but almost everyone called it a freezer (a term still used today). From the freezer, the then heat-laden vapor went to a device which, by "squeezing" reduced the vapor in volume. This "compression," as it is termed, took place in a cylinder that was closed at one end with a piston moving up and down in it. A crankshaft and connecting rod moves the piston (see Fig. 1-3). The crankshaft had a pulley on one end,

Fig. 1-3 The refrigerant cycle

connected to an electric motor by a belt. This "up-and-down" movement, or "reciprocating action," led to naming the device a "reciprocating compressor."

The flow of refrigerant in and out of the cylinder was governed by "one-way valves." In Fig. 1-4 we see how the valves, which are flat pieces of spring steel, work. When the piston in the compressor shaft goes down, the inlet valve bends to allow vapor to be drawn in, and, when the piston goes up, the inlet valve bends back to cover the inlet. The vapor then will be pushed out at higher pressure through the outlet, and the outlet valve will then cover the opening to prevent the vapor from returning through it. The cycle is repeated.

Because compression made the vapor hot, and this heat must be removed in order for the vapor to return to the liquid state, a method had to be devised to remove this heat of compression. The vapor flowed through some tubing, which in turn had air flowing over it. This air cooled the hot vapors. Inside the tubing, the cooled vapor formed drops of liquid that trickled to the bottom (see Fig. 1-5). The compressor pushed the vapor into the tubing so fast that a certain amount of pressure was maintained—enough, in fact, to cause the liquid to run into the freezer to repeat its cycle. This process of a vapor changing to a liquid because of cooling is called "condensation," and the part of a refrigerator where it takes place is called a "condenser."

In the old refrigerators, the freezer, compressor and condenser, and the tubes connecting them, were put together with bolts and screw-on fit-

Fig. 1-4   Operation of the "one-way valves"

tings. This arrangement is called an "open" system, as distinguished from a modern "sealed" system, where the connections are soldered, or "sealed," so that the refrigerant cannot get out and air cannot get in.

We will not devote much space to the study of open reciprocating mechanisms because, at least for household use, they are rapidly disappearing from the American scene. Sulphur dioxide is corrosive and smells horribly (like rotten eggs) when it escapes. It served its purpose at the time, however.

In the so-called open units, the flow of refrigerant is governed by a float valve. One type of float-valve assembly consists of a tank with two openings. The larger opening is a pipe through which liquid refrigerant moves freely (see Fig. 1-6). Inside the tank there is a float that rises and falls as the liquid rises and falls. When the float drops, the leverage action of an arm attached to it causes a needle point to move away from its valve seat, opening a small hole through which liquid could flow into the tank. As the float rises, the procedure is reversed with the reseating of the needle point, preventing more liquid from running in until called for by another drop in the liquid level.

All float valves can develop trouble. If the float leaks and fills with

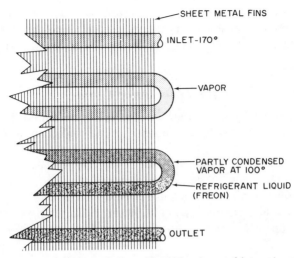

Fig. 1-5 This condenser diagram illustrates how refrigerant vapor condenses to a liquid when cooled

Fig. 1-6 The float valve

liquid, the needle point will be drawn to the end of its travel. Or, the point might adhere to its seat and not open at all. If a particle of metallic dust lodges between the needle point and the seat, it will not close properly. Needless to say, these controls were a source of considerable trouble.

One of the early refrigerators was called a "low side" system. The freezer had a tank (with float valve) arranged so that the liquid could run into the sheet metal (or perhaps cast metal) passages of the freezer. The term "low side" was coined because the pressure of the vapor in the freezer was low compared to the pressure of the entering liquid. The needle point may be thought of as the dividing line between the low and high pressure parts of the system. This idea is worked out in all forms of refrigerating equipment; its present day application will be discussed later.

Thus, we may visualize the refrigerant as running into the freezer where it evaporates, absorbing heat from the surrounding air (through the freezer walls), and then being drawn (as heat-laden vapor) to the compressor.

Why the Hot Vapor Condenses

We mentioned that a vapor will condense to a liquid if it is cooled; that is, it will *eventually* condense if it is cooled to a low *enough* temperature.

In the case of the sulphur-dioxide refrigerant, if the amount of vapor in the condenser was such that the pressure of the vapor was the same as that of the outside air, the condenser would have to cool the vapor to at least 14 °F. to make it become a liquid. Because that temperature is way below room temperature, we might ask, "Why *does* it go back to the liquid state?"

The answer to this question is found in a previous statement that "a certain amount of pressure is maintained." It is a principle of physics that the more you compress a vapor, that is to say, the more you raise its pressure, the higher you raise the temperature at which the vapor will liquify or condense. In the case of sulphur dioxide, by having the vapor at a pressure of 60 pounds per square inch (psi) or more, enables it to condense at 75 or even 90 °F.

The reason that the pressure in the condenser can build up so high is that the float valve holds back the refrigerant and lets it run into the freezer very slowly. Thus, the float valve serves two functions: on the one hand it lets the liquid go into the freezer only as fast as it can evaporate, and on the other it allows pressure to build up in the condenser.

Another early type was the "high side" system. The float was in a "receiver" tank underneath the condenser. In this case, the float let liquid *out* when the float was raised. The control permitted just enough liquid refrigerant to keep the pipe or tube to the freezer (the "liquid line") and the freezer itself filled. With this arrangement, small bubbles of vapor could form on the freezer walls and rise to the top to be drawn off. Whether this happens as shown in Fig. 1-7A, or whether a drop of liquid runs onto a flat surface and completely evaporates, as shown in Fig. 1-7B, the result is the same: heat is absorbed.

Freon—The Modern Refrigerant

In the sulphur-dioxide era, much research was done to devise a substitute for the foul-smelling gas. When methyl chloride became available, it was viewed as a blessing. It (normally) maintained a positive pressure in all parts of the system and did not smell badly when it leaked. Its popularity was brief, however, as it lost favor when the DuPont chemists invented the

Fig. 1-7 An evaporating liquid absorbs heat

"Freons," which nearly all household units use today. We shall refer to modern refrigerant as Freon, although the Allied Chemical Company (and some other concerns) are making a similar product with a different name.

There are several varieties of Freon, differing in their vaporizing temperature. Freon-22, for instance, boils at about 41 °F. below zero. It is difficult to think of anything *boiling* at such a low temperature; however, it is not uncommon for most gases. Freon-113 boils at 117 °F. above zero and Freon-11 at 75 °F. above. We can conclude that in a room at 76 °F. it would be easy to liquify Freon-11, and that in a room at 74 °F. it would be easy to boil it. It is a very complex matter, however, to apply these ideas to the actual design of a refrigerator and to say precisely what degree of cold will be obtained with a given set of conditions. The internal pressures determine performance, and you will have a different pressure at every inch of the circuit; all the bends and restrictions will change the pressure. Furthermore, you will have the radiated heat of the motor to contend with, as well as the circulation of air, which might be either warm or cool. As a matter of fact, the computations are so involved that rapid refrigeration research had to await the development of the electronic computer.

Freon-12 is used in nearly all household refrigerators and some air conditioners. The next most common type is Freon-22, which is widely used for air conditioning and commercial refrigeration. "Freon" is often abbreviated, "Freon-12," for example, appearing simply as "F-12." When other companies began making these chemicals, they were not allowed to use Du-Pont's term; consequently, they called them "refrigerants" and preceded the number with an "R." F-12 and R-12 thus refer to the same substance. It

Fig. 1-8   A 115-volt basic circuit

is always necessary to have the number correct, whatever the preceding letter may be.

Some technicians claim that if you mix refrigerants, there is a possibility of a reaction taking place that will create a great deal more pressure than either refrigerant would by itself.

For our purposes, it will suffice to say that a unit is supposed to have a particular type of Freon in it, and, if a different type is used, the unit will perform poorly or not at all.

Modern Controls

With the advent of the sealed unit, a change was made in the starter switch. The belt-connected motor had a centrifugal switch that disconnected the starting winding when the motor speeded up enough to move a weight away from the shaft. Obviously, in a welded motor-compressor housing you would not be able to get at it for repair. The present switch, mounted externally for this reason, is called a "relay." In addition, the troublesome float valve was replaced with a "restrictor," of which we will learn more later. The operation of the sealed unit is regulated by a thermostatic switch, or "cold control."

The sealed unit, relay, and cold control are the basic working parts of the conventional refrigerator (see Fig. 1-8). Engineers have been able to take these "basic parts" and somewhat modify them and make new devices for other purposes. One of the most important of these is the air conditioner.

New Developments

In the matter of air conditioner economy, a style of compressor has been built that features a two-speed motor. Industry sources say that a reduced capacity for mild weather may be achieved by putting two motors in one shell. The making of a larger shell for two production-line motors is cheaper than putting special windings in a single motor. In 1979 the Swedish Stahl screw compressor was introduced. Large capacity variation and reliability are claimed for it.

Some buildings are now being designed using every form of energy conservation the mind can devise. Of the many ideas, three are simple: the berm, the trombe wall, and ventilation.

As for the first, in the old days people knew that shading the windows was a great help in surviving the heat. Store windows had roller awnings, and homes had attractive striped window awnings that could be raised and lowered. They also had another form of shade—*trees*. In their zeal to clear the land, developers have been cutting them down. A *berm* is a big pile of dirt meant to create shade on the *sunny* side of a building.

A *trombe wall* is a wall that faces south and has a lot of windows with drapes or venetian blinds. When the sun shines in, it warms a flagstone floor that gives off its heat during the night; for cooling, the floor is chilled by the night air and the windows are shaded during the day.

The third energy conserver is ventilation. Some parts of the country undergo a drastic change in temperature when the sun sets. A fault of insulated houses is that they stay warm when the outside temperature drops. Running a compressor is inefficient when the outside air is cooler. The power consumed by a separate fan to pull in the outside air would be trivial in comparison. The most reliable control mechanism for this purpose is the human hand. But a timer can be used to turn on a fan at, say, three in the morning to fill the attic space and the rest of the house with cool air. That would keep the air conditioner off for a few hours in the morning. The solar heat people would be glad to tell you how to put cinder blocks in an insulated box to store cold; using them would allow you to keep the unit off for a few additional hours.

As for solar cooling, in 1975 some Israelis devised a way to use a parabolic reflector to create a heat higher than the 300°F required for an absorption system. Then American researchers found absorption chemicals that would work at 200°F. Because this is closer to a possible air temperature of 115°F, it is more practical.

In the thermoelectric module, a small "interface" has a pair of wires attached. When energized, the plate will get cold. This "Peletier Effect" has been known for a long time, and some small "ice boxes" make use of it. The interfaces are expensive, however, which explains why we do not see the idea employed on a larger scale.

In the matter of the storage of heat or cold, a promising development is the use of certain chemicals that go between the solid and liquid states at room temperature. One method is to store material in plastic bags with a substance that solidifies at night and absorbs heat when it liquefies during the day.

Some new designs for "total comfort" systems feature a multitude of sensors connected to a "microprocessor" box containing printed circuits and solid state devices. Servicing these complex arrangements should be a good line of work for anyone who is willing to undertake the considerable training required to cope with them.

2

A Look at Air Conditioning

Keeping cool in the summer must have been one of man's earliest endeavors. The Babylonians had their own version of how to reduce the temperature in a home by running cool water in stone conduits under the floors. (They also warmed the water for radiant heat in the winter!) From then until Thomas Edison died while telling his listeners how to air condition the room he was in, no one thought of a practical device to combat the summer heat.

The present-day home equipment is a byproduct of temperature and humidity control, and air filtering developed for industry. In the beginning, comfort was only incidental. But, it would be quite proper to say that the modern home air conditioner is a "comfort-making machine." It has several functions that contribute to this end. The first, obviously, is cooling the air, which conserves body energy by lightening the load on the circulatory system.

The second function is dehumidifying the air. We mentioned that warm air can absorb and hold more moisture than cool air. But, without knowing the statistics, it is difficult to realize how very sharply the moisture holding capacity of the air increases as the temperature goes up. A pound of water weighs 7,000 grains. At 70° F. a pound of air will hold 110 grains of water. But the figure jumps to 156 at 80° F., 218 at 90° F., and 302 grains at 100° F. Thus, with 100% humidity you would have almost three times as much moisture in the air at 100 as at 70° F. Perspiration cannot evaporate as readily when the air is already moisture laden, and this produces the well-known "sticky" and uncomfortable feeling.

Another fact of nature is that rainstorms occur when a cold front collides with a moisture-laden air mass—simply because the water vapor condenses to rain when the air is cooled. Applying this principle to air conditioning: when the moist hot air passes through the cold evaporator, some of its moisture is deposited.

The third function of the device is to filter the air. This makes the house easier to keep clean, of course, and it makes for healthier living conditions by removing, or at least reducing, air-borne pollen and irritating industrial dust. This has proven a boon to persons suffering from respiratory diseases.

A fourth function is to circulate the air, breaking up stagnant "pockets" and bringing about an even temperature in the various parts of the room.

Some units perform a fifth function—providing heat when the temperature drops below a specified point.

Types of Air Conditioners

There are several types of home air-conditioning units. The earliest was the window type, which was readily adapted for "through-the-wall" installation. There also is the "free-standing" type, which stands away from the walls and windows and takes in, conditions, and expels the air without benefit of any ductwork for either the air supply or distribution. These are used mostly in places of business that have large rooms to take advantage of the relatively large capacity of these units.

The latest development, and one that is finding wide acceptance throughout the country, is "central" air conditioning, in which cooling coils are placed in forced-air ducts so that the cool air can be distributed throughout the house. There are several methods of installation. The location of the cooling unit itself, or parts of it, and the size and placement of the ducts is influenced by the design and structure of the building. It is a matter of determining what installation is compatible with economical operation of a system that will provide efficient cooling. This is a complex matter that has been given considerable study.

In new construction, several factors are given more consideration than they were in the pre-air conditioning days. To quote Carrier's "Home Management" booklet:

"Principles relating to utilization and control of heat from the sun become extremely important in an air conditioned home. Careful attention should be given to exterior design, orientation, size and placement of windows, colors, insulation, and proper selection and installation of air conditioning equipment. By following this approach, one builder saved one-third in first cost of air conditioning and one-half of annual operating cost in a house of medium price."

We have said that the basic principles of operation of the "comfort machine" are the same as that of the refrigerator described earlier. The principle parts of the air conditioner, however, are not necessarily all in the one cabinet as they are in a refrigerator. In Fig. 2-1 we see the compressor and condenser outdoors and the refrigerant lines extending through the wall to reach the evaporator, or cooling coil, which is in the duct system of a furnace. The warm air is drawn into the unit, over the cooling coils, and distributed over the ductwork.

Fig. 2-1 Furnace with separate condensing unit (Reprinted by permission of Carrier Corporation. ©1979 Carrier Corporation)

In Fig. 2-2 we have a self-contained unit outdoors on a "pad" and connected by a collar to the sheetmetal box on the top of the furnace. Here the return air is drawn through the furnace filter and up into the cooling unit, where it is reversed in direction and pushed out into the ducts. The flow of air is governed, of course, by baffles that are not shown.

The cooling unit can be either electric or gas fired. In Fig. 2-3 we have gas used for both heating and cooling. Cooling with heat may seem inconsistent, but Chapter 9 explains how it is done. In this case, the cooling unit chills water, which is forced through pipes, cooling the duct coils as it circulates.

In Fig. 2-4 we have another system which uses water, but, in this case, the water cools the condenser instead of being cooled by the evaporator. In the process, the water picks up heat, which it then must lose in a "cooling tower." The one shown rests on the ground, and uses air to remove the heat from the water. Of course, circumstances might dictate that it be placed on the roof.

Fig. 2-5 shows methods of air distribution. We have, in Fig. 2-5A, the unit in the basement with ducts running under the floors. Figure 2-5B shows the unit in the center of the house with ducts running through the cement-floor slab.

Fig. 2-2 Furnace with self-contained air conditioner (Reprinted by permission of Carrier Corporation. ©1979 Carrier Corporation)

Fig. 2-3 Two-piece gas-fired cooling unit (Reprinted by permission of Carrier Corporation. ©1979 Carrier Corporation)

Fig. 2-4 One-piece gas-fired cooling unit

Fig. 2-5 (A) Perimeter system in basement of house; (B) perimeter system in house built on foundation; (C) center duct system; (D) attic system with ceiling registers (Reprinted by permission of Carrier Corporation. ®1979 Carrier Corporation)

A perimeter system puts the cool air where it can diffuse the heat of the outside walls, thereby creating an even temperature. In Fig. 2-5C we have a center duct running *under* the ceiling. Figure 2-5D shows a method for installing a *second floor* cooling system with little alteration of the existing structure. The distribution ducts extend sideways from the center, or "trunk" duct, and direct the air downward through ceiling openings. In all of these systems, return registers and ducts are needed but are not shown.

Perimeter Vs. Central Duct System

Whether a perimeter system is used will be dictated by the severeness of the winters. With extreme cold, it is desirable to counteract the chill of the outside walls, making a perimeter system almost mandatory. If the house has a forced-hot-air system, the most economical method of installing central cooling is to utilize the supply and return ducts that are provided.

In any case, the air circulation in the conditioned spaces is governed by the shape, size, and location of the registers, as well as the capacity of the fans and ducts. How the air circulates within a room can be determined by holding a cigarette in the air stream and observing the direction in which the smoke travels.

A combination heating and cooling plant is termed a "year-round air conditioner." Where the furnace is of the steam or hot-water type, the air-conditioning system would have to be installed separately.

The foregoing does not exhaust the possible methods of installing central air conditioning; the methods mentioned are merely the most commonly used.

The Window Air Conditioner

We already know that refrigeration is the heart of the air conditioner. Now, what we want to learn is how refrigeration is utilized and what might go wrong if the unit does not perform properly. The easiest way is to study a small unit first. An early-type window air conditioner with a simple "on-off" switch is ideal for this purpose; other than the switch and the fans, the only other parts are the sealed unit and a starter for the motor.

Because the condenser gives off heat when the unit is in operation, the first consideration is, obviously, to expel the heat from the conditioned room. It would be quite pointless to allow it to return into the room that you were attempting to cool. Removing the heat is accomplished by using two fans—one blowing air through the cold evaporator, and the other moving air over the motors through the hot condenser and removing it from the conditioned room. Figure 2-6 shows a typical arrangement in such a unit, and Fig. 2-7 shows an actual window unit with the parts labeled.

We have now considered two conditions that produce vaporization—

Fig. 2-6   Air flow and condensed water removal in window unit

Fig. 2-7   Components of a window unit (*Courtesy*, Frigidaire)

that which takes place in the freezer compartment of a refrigerator, and that which takes place in the finned tubes of an air conditioner evaporator. A great deal of research has been devoted to the latter method of vaporization, especially in the use of a similar device to cool automobile engines. The modern finned evaporator (or condenser) is able to transfer a considerable amount of heat for its size.

Your unit might have a fan motor with a fan on each end of the shaft, or it might have a separate motor for each fan. In the latter case, the two motor casings would probably be bolted together to save space, but this would not affect their operation (see Fig. 2-8).

EVAPORATOR   CONDENSER

Fig. 2-8 Fan motors bolted together (*Courtesy*, General Electric Company)

The conventional arrangement has a "cage" fan for the room air. It draws in filtered air through its center opening and throws it to the periphery where it strikes a baffle and is reversed in direction so that it blows into the room after passing over the cool tubes and fins. Fresh air from outdoors may or may not be introduced to the air stream on its way to the tubes. A cage-type fan is able to develop the thrust necessary to pass the air over the tubes, through the filter, and still have it travel into the room with some force. A cage-type blower wheel is shown in Fig. 2-9.

With a large volume of air passing through the evaporator, it follows that considerable moisture will be deposited on it. On the early, small models, the moisture (or condensate) was disposed of simply by letting it drip down into some troughs that returned it to a pan under the condenser. At that point, the heat and turbulence of the air would evaporate it. If it overflowed the pan and dripped onto a lawn, no harm would be done. Dust or traces of oil might get into the water, however, so that if the wind blew it onto the building an unsightly stain might result. A more positive method of dealing with the condensate was indicated. The newer models have a "slinger ring," which is a ring or rim that attaches to the fan blades. It spatters or "slings" the water onto the hot condenser (see Fig. 2-10).

Another development introduced switches for various fan speeds and various degrees of cooling. An electric heater, wired so that it would run with the compressor shut off was also added, but because the heater coils are of limited size, it should not be expected to do much except take off the chill in the spring or fall (see Fig. 2-11).

FRESH AIR EXHAUST   FAN MOTOR   MOTOR BLOWER
INLET   DOOR   MOUNTING BOLTS   SHAFT WHEEL

Fig. 2-9   Cage-type blower wheel (*Courtesy*, Frigidaire)

The Heat Pump

The next step was to develop a device so that the unit could furnish heat on chilly days without requiring red-hot heater wires or the like. It has been said that a customer wrote to a manufacturer, saying that his window air conditioner had accidentally been installed backwards and he found he could use it for a heater. That is not the way it happened, but the statement has a degree of logic because the unit is, so to speak, hot at one end and cold at the other.

More than a hundred years ago an Englishman, Lord Kelvin, did the paper work on a device to take whatever heat could be found in air or water and transfer it to a warmer medium. In other words, to draw heat from air at, say, 50° F. and use it to maintain a room temperature of 70° F. A Mr. Kramer, in the modern era, was able to construct such a machine. Since then, scientists have theorized many uses for so-called heat pumps, such as

FRESH AIR DOOR   FRESH AIR DOOR SPRING   FIBERGLASS SLEEVE

CONDENSER FAN   CONDENSER SHROUD   FAN MOTOR   PLASTIC CENTER PARTITION   MOTOR PROTECTOR ASSEMBLY

Fig. 2-10   Another view of center section (*Courtesy*, Frigidaire)

HEATER TERMINALS   FUSE LINK   SAFETY THERMOSTAT   HEATER ASSEMBLY MOUNTING TABS

HEATER RECTIFIER ASSEMBLY   HEATER ASSEMBLY   HEATER ASSEMBLY MOUNTING SCREWS

Fig. 2-11   Heater assembly

drawing water from under ice and extracting heat from it and using that to heat a house. The impractical aspect of that scheme, however, is that you have to do so much pumping of water and refrigerant that the power cost would be higher than the value of the heat obtained.

But, actually, the same idea is used in your refrigerator. In the freezer compartment, the temperature is below freezing, but the unit takes heat that seeps in and expels it to the much warmer room through the condenser. It would follow that, by revolving the window air conditioner, it could be used as a heater on a cool day. It is impractical, however, to have a "merry-go-round" arrangement so the user can turn the air conditioner as desired.

Faced with this, the designers devised a way to make the condenser act as the evaporator, and vice versa. In the window air conditioner the two coil banks are similar in construction to make this substitution possible, and, with the use of a "reversing switch," the desired action was accomplished.

Today, industry is trying to eliminate the use of seals and packings on external connections. When you start putting valve stems and the like on a sealed unit, it is no longer sealed, and, with respect to leaks, it reverts to the obsolete "open" system. The problem was to devise a control for the refrigerant. This was solved by the use of a solenoid (see Fig. 2-12).

Fig. 2-12 Operation of solenoid

A solenoid has an electric coil, which, when energized creates a magnetic field inside the solenoid. The field extends into a nonmagnetic casing (such as brass). The force developed by the magnetic field moves a plunger inside of the casing. The plunger is spring loaded, that is, there is a coil spring inside of the casing that rests against one end of the plunger. When the coil is energized, the plunger compresses the spring, and when the current is shut off, the spring returns the plunger to its original position. Energizing the coil diverts the flow of a small stream of refrigerant, which moves a piston connected to a valve.

Why not use a solenoid to move the valve directly? Well, that could be done, but, with this other arrangement, the power of the motor and the pressure of the gas (refrigerant) can be used. Consider what happens when you raise a car with a hydraulic bumper jack. Moving the handle pushes a small piston that pumps oil from its chamber into a larger chamber with a larger piston. If you pump oil into a chamber ten times as large, the piston in this chamber moves only one-tenth the distance as the smaller piston. It works like a reduction gear—you trade speed for power (see Fig. 2-13).

Fig. 2-13   A hydraulic pump

In using the solenoid system, the small refrigerant stream governed by the solenoid will go into, say, the left end of a cylinder which has a piston with a reversing valve (see Fig. 2-14). More likely there are two pistons with the valve, mounted on the arm between them, sliding back-and-forth as a single unit. With the solenoid letting gas into one end of the cylinder (1), the valve assembly would slide to the left, and conversely, when the gas flow was switched to the other end (2), the assembly would slide to the right.

The valve itself is simply a plate with a channel cut out. This plate slides on another plate with three holes, or "ports," that are in line and, at

Fig. 2-14 (A) Heat-pump cooling circuit, and (B) heat-pump heating circuit (*Courtesy*, Frigidaire)

any time, two of them are covered by the channel in the valve plate. The
center port, "B," receives the hot, highly compressed gas from the compressor. Port "A" leads to the indoor tubing and "C" to the outdoor tubing. With the valve plate slid all the way to the right (Fig. 2-14B), the hot gas
would be directed into the outdoor tubing, and similarly, with the valve
plate to the left, the hot gas would go to the indoor tubing.

There is no question that this arrangement can provide warmth, but it
is not a substitute for a heavy-duty heating plant.

Various companies have different notions about the automatic control
of the hot-gas and electric features. Some newer models have two electric

resistance heaters, and the "program" for control might go like this: With the switch on "heat," if the room temperature drops 2° below the thermostat setting, a 2200-watt heater will be energized. If it continues to drop, the compressor and reversing valve will be energized. If the outside temperature drops to 46° F., the compressor will shut off and the second heater will go on. (That requires another thermostat outdoors, wired so that it can break the circuit to the compressor.)

The Defrost Heater

When these units were first operated, it was found that considerable frost deposited on what was, for cooling purposes, the evaporator. That called for a defroster. If the model had a radiant room heater, that was utilized as the defroster. If not, defrosting was accomplished with a radiant (red-hot wire) heater of smaller size. The defroster control switch was either on the suction line (see Fig. 2-15) or on the evaporator (see Fig. 2-16).

A window location was inconvenient for some people, resulting in the manufacture of through-the-wall air conditioners. In new buildings, many contractors provide space for this type of installation.

Before you "do" anything to a window air conditioner, carefully study the mounting plates, brackets, and fixtures. In some cases, these

Fig. 2-15 Defrost control mounted on suction line (*Courtesy*, Frigidaire)

Fig. 2-16   Defrost control mounted on evaporator (*Courtesy*, Frigidaire)

metal parts have corroded to a point where they can "give way" if the unit is subjected to slight movement. By shifting the unit to remove panels or parts, you might break the mounting fixtures, creating a hazardous situation.

Some metals are susceptible to corrosion at any location, and in some sections of the country corrosion proceeds very rapidly. Industrial fumes contribute to corrosion, but the main problem is salt carried by the air. In certain coastal areas, the deterioration of mounting parts is so rapid that stainless-steel parts eventually might have to be used for safe mounting. Aside from the corrosion problem, some window units are not very securely installed in the first place.

Improper Cooling

When a window conditioner does not cool properly, there are several possible reasons *not* related to the sealed unit or the controls. It is surprising how often a repairman is summoned only to find that drapes or venetian blinds are covering the front of the unit, thereby restricting the flow of air. And there are some other causes of poor cooling that are neither mechanical nor electrical. For instance, the condenser can become covered with insects. Certain insects can cause a great deal of trouble because, in cool weather, they will look for a warm area. They can get inside the cabinet, and, when the unit starts, the fan will blow many of the insects onto the tubing and fins. Most of them can be removed with a vacuum cleaner.

Another cause of poor performance might be a filter that is covered with dust. Or maybe the filter has been accidentally left out of the unit, thereby allowing the evaporator to become covered with dust. Many types of filters are in use. There is the old disposable kind, of course, and some plastic and pressed-metal types that, it is said, can be washed and re-used. There is a sponge-rubber filter that, the manufacturer claims, will "arrest the movement of pollen and bring heaven on earth to hayfever sufferers."

The admixture of outdoor air is governed by a baffle. Although the baffle is supposedly rugged, it could warp, thereby allowing hot air to come in at a time when it is not wanted. Perhaps the linkage between the baffle and the knob is damaged. Or the knob might possibly have turned on the shaft so that, in the "recirculate" position, the baffle does not close. For that matter, the fill-in panels between the unit sides and window casing might be loose, allowing hot outside air to enter. Or it might be that the unit simply is not big enough. In this case, the fact will not become apparent until a heat wave comes along.

Water Leaks

Water leaks are another concern. With high humidity, the water removed from the air might be considerable—as much as a few pints per hour. If something is obstructing the flow of water to a degree that the slinger ring cannot pick it up, the water might drip on the floor. Or, if the front of the unit has dropped, this would prevent the flow of the water in the proper direction. If water does drip from the back, the slinger ring might be out of order, or it might be a normal condition caused by high humidity.

Noise

A certain amount of noise is normal in an air conditioner. The Freon will make a hissing sound as it travels through the system. If you pulled the plug you might hear it until the high side, or "head" pressure, had diminished through the restrictor. When operating, the "pulsing" of the compressor can set up a harmonic vibration in the tubing. If the vibrating tubing rubs on something, a "rattling" noise would ensue. (The tubing can probably be bent to stop the noise. If not, a piece of sponge rubber can be inserted to absorb the vibration.) A "thumping" noise heard when the motor runs might be traced to the motor mounts (see Fig. 2-17). Three things might be wrong: (a) some motor-compressors are shipped with

Fig. 2-17 Typical motor mounts

special bolts that should be removed and discarded—they might still be in place; (b) another type of mount has a spring and/or rubber arrangement with a bolt that should be loosened and left in place—the bolt might not be loosened enough—this is easily checked by pulling the plug and wiggling the motor; (c) in another compressor mount rubber blocks are inserted to absorb the vibration and hold the noise down to a reasonable level—here again, it might be necessary to turn the bolts that hold the mounts in place. Sometimes a little experimentation is required.

If the air stream increases greatly in velocity and its attendant noise becomes objectionable, the trouble might be excessively high voltage making the fans run faster. That is a rare occurrence, but, if it ever happens, the power company should be notified at once.

Ticking noises are caused by the fan blades, but they can be bent so they will not hit anything. Scraping noises might be caused by the slinger ring. Here again, bending or adjusting will provide clearance.

A definite "squeal" might be caused by a dry bearing. If you do not oil it, the whole unit might become inoperative. When the bearing finally seizes, it will stall the motor. If the motor does not have an overload protector, the excessive draw of power will blow the fuse.

Mention should be made that in a condenser or evaporator, the term "coil" is used even though the tubing is not spiraled but just goes back and forth in loops. (You might find a spiral of tubing if it were in a tank of water). As far as window air conditioners are concerned, the term is a convenient one to use.

A "pin" coil is presently in use on which the fins have been replaced by a multitude of welded pins. The heat transfer of this type is increased 50%, but a cleaning problem would be encountered if the filter were left out of the unit. Dry dust would not be much of a bother, but certain vapors, especially kitchen grease, will quickly form a sticky film on the metal, and "seize" whatever dust comes along.

Some air conditioners have been fitted with a timer, the rationale being that the cooling can be shut off when there is nobody at home and the timer set to go on two or three hours before an anticipated return (see the control panel shown in Fig. 2-18). Others may have a built-in low voltage cut-out, which will cause the timer to reset and recycle after a few minutes. To keep the fan from running all the time, a device called "comfort selector" (or some such term) will shut it off when a room is cool and keep it off until the temperature has risen to a predetermined point. Use of such a device does save money but also causes a greater variation in temperature. There is also an arrangement to oscillate the vanes to give a better distribution of air in the room.

Fig. 2-18   A control panel, with timer

With no circulation, the air will "stratify" in layers, with that nearest the ceiling being quite hot and that nearest the floor quite cool. The blades of an old-fashioned ceiling fan, which revolve at a level somewhat below that of the ceiling, can be used to set up a circulation pattern that pushes moderately cool air down to the floor and entrains the much cooler air there so that it moves across the floor and up the walls and then across the back to the place of origin. This movement helps evaporate perspiration. Some people claim, in fact, that a fan of this sort makes it feel like 70 when the thermostat is set at 78. Another advantage is that a large-blade, slow-moving fan is quite economical to operate.

An interesting new development is the electronic air filter, an example of which is shown in Fig. 2-19. The air that passes through it is ionized by the wires for this purpose that are shown in Fig. 2-20.

Split systems offer several possibilities when pipes are passed in through an outside wall. Since pipes can be extended to any space in a house, evaporators can be placed in closets or recessed in an interior wall.

Fig. 2-19 Electronic air cleaner assembly (*Courtesy*, Frigidaire)

Fig. 2-20 Air cleaner cell (*Courtesy*, Frigidaire)

Evaporator coils have often been installed in the ductwork of an existing hot air furnace.

Mobile home installations require special handling because mobile homes are long and narrow, or at least the old ones are, and additional velocity is needed to push the air through the ducts. Even the new double-width homes may not have their ducts arranged so the air will flow as readily as it would in a house. A "velocity gauge" will be helpful here. Since mobile home units usually have a variable-speed fan, fan speed can be set as appropriate. This type of unit would probably have a crankcase heater (an electric resistance wire to warm the oil) and a rather large electric heater for supplementary heat. The 20-kW size uses 70-amp fuses. To limit the in-rush, one arrangement increases the power to the unit in stages. Heaters can have fusible links as well as thermal limit switches.

In all cooling equipment, the collection of frost and ice on the evaporator coil (assuming there is enough refrigerant) depends on the amount of hot air passing over the coil. Dirt on the coil or filter or a foreign object can block the air flow. It should be noted, too, that mechanical problems can slow down the air delivery of a belt-connected blower. The belt can slip in the same way it does in your car. The frost-ice problem is often remedied by having the thermostat bulb touch (or come close to) the coil. As soon as ice touches the bulb, the switch will "think" the room is cool enough and shut off the unit. Exceptions are those instances when the outside temperature drops so low that it "shrinks" the volume of the refrigerant and when low-voltage conditions slow down the fans.

Heat pumps have a separate defrost control (see Fig. 2-21). The fittings are for high- and low-pressure tubes that extend into the coil. Robert-

Fig. 2-21   Defrost control (*Courtesy*, Robertshaw Controls Co.)

shaw Controls Co. calls this a simplified control—a "demand" type with a thermostatic bulb inserted in a "well" on the coil. The bulb automatically decides when frost needs to be removed and causes the unit to revert to normal operation when it has been.

Ice on an evaporator tends to pass liquid refrigerant into the compressor. Manufacturers have to provide for this contingency with a thermal shut-off or an accumulator (a sort of tank to hold the surplus).

An interesting development for the window type of unit is a drain water control for the pan (see Fig. 2-22). In warm weather, it holds the water in the pan to be available for splashing on the condenser, thereby improving its efficiency. When the temperature drops to near freezing, it lets the water out.

Fig. 2-22   Drain water control (*Courtesy*, Robertshaw Controls Co.)

Evaporators are now being fitted with a fan that offers different speeds, the rationale being that people can often get along as well, or better, without constant blasts of cold air from their units.

Evaporators do not cool as well when they are wet. This is another way of saying that some percentage of the power input has to be used to condense the moisture in the air. In one unit, when it is 100 °F outdoors, the indoor air will drop about 14 degrees on one pass through the evaporator at 50-percent humidity and about 13 degrees at 60-percent humidity. These figures make it clear that there would be an appreciable difference in efficiency between a unit in New Orleans, say, with that city's very high summer humidity, and one in the Western deserts, with their dry air. The figures given are for a room temperature of 75 °F. In a warmer room, there will be a greater differential between the room air and the evaporator temperature and consequently more cooling each time the air passes through.

A measurement called the Energy Efficiency Ratio (EER) provides a basis of comparison between various models. This ratio is arrived at by dividing the output in Btu by the input in watts. The figure thus obtained tells how many Btu of cooling can be obtained with a watt of power. A high efficiency rating requires (1) an EER of 7.5 or better, or (2) a better than midpoint rating in the Department of Commerce chart that lists the products of various companies in each size classification.

Figure 2-23 shows the various wall receptacles now available. Figure 2-24 shows a motor protector with a three-pin terminal. Figure 2-25 shows the orientation of the compressor pin terminals used by several companies.

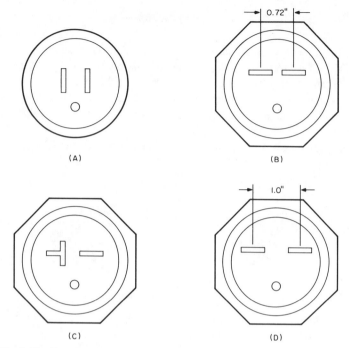

(A)

(B) 0.72"

(C)

(D) 1.0"

Fig. 2-23 Two-pole, three-wire grounding receptacles: (A) 15 A, 125 V; (B) 15 A, 250 V; (C) 20 A, 250 V; (D) 30 A, 250 V

TERMINAL COVER GASKET   MOTOR PROTECTOR   PROTECTOR MOUNTING SPRING

Fig. 2-24 Motor protector and terminal pins

Fig. 2-25 Compressor pin terminals

Figure 2-26 shows a mobile home installation with flexible ducts running to the main distribution ducts and the return air register. Figure 2-27 shows a piston compressor.

Considering the heavy amperage draw of air conditioners, it is worthwhile to consider the hazards of aluminum house wiring. Since the expansion rates for aluminum and copper are different, arcing can take place at junctions of the two metals and little globules of molten metal may fly out. After many such instances contractors learned to use a special kind of grease at the points of contact.

A popular testing device is the "clamp-on" volts-ohms-ammeter, which gets its name from two metal "fingers" that can be opened and put over a wire for an amp reading. When you take your thumb off the button, the fingers come together again. Whenever a flow of ac current reverses,

Fig. 2-26   Mobile home installation

Fig. 2-27   Piston compressor (*Courtesy*, Bristol Compressors)

some electrons are ejected from the wire and caught in the insulation, causing a slight "emanation" in the air that a sensitive instrument like this one can sense and record on a dial.

Use of any volts-ohmmeter presents the possibility of testing for volts when the device is set for ohms. One of the old types would be ruined by this misstep. New types are usually equipped with a fuse as a safety measure.

Condensers can be cooled by air or water. The temperature of tap water remains fairly constant the year around, at least much more so than that of air. The wide variations of the latter do not contribute to efficient operation. Extreme heat can cause too much head pressure and extreme cold, not enough. The flow of water, however, can be regulated to keep the condenser at an appropriate temperature. In 1979, nearly all units sold were of the air-cooled type, perhaps because many buyers were afraid they could not cope with scale removal. Water quality is so variable that mineral content can differ in different parts of the same township. The people who make descaling chemicals are the ones to contact when scale problems arise.

The heavier equipment runs on 208/230 V, but it is easier to learn the principles by studying the 115-V units first. The next three chapters will deal for the most part with the difficulties of a 115-V air conditioner with thermostatic control.

3

Unit Inoperative

The unit will not run, obviously, if it is not receiving electricity. The fuse might have blown or the circuit breaker might have kicked off. If the fuse has blown it does not necessarily mean that anything serious is wrong. Fuses sometimes get "old and tired" and blow for no other particular reason. With either type of safety device, if there are several appliances on the same circuit, and they all started up at the same time, something might "give." A motor needs more current to start than to run so that, if, by coincidence, they all started together, such an abnormal load could blow the fuse. Or, they might all run for years without that happening. If the power supply should fail for a time because of a storm or repair work being done on the lines, all the thermostatically operated devices might start at the same time, as well as anything else (such as an appliance) that happened to be in use when the power failed. In this situation, the excessive (and abnormal) power demand would most likely blow the fuses. When this happens shut off the appliances, replace the fuses, and start the appliances one by one.

If the equipment is old and has a two-prong plug, the problem may simply be that the plug has fallen out of the receptacle or that the prongs are not making contact. (If a plug fits loosely and the cord is "jiggled," one or both of the prongs may lose contact.) If the plug is in place, remove it, pull the prongs apart, and reinsert it.

Use a Tester

To find out if an outlet is energized and to troubleshoot beyond that point, it is necessary to use a test bulb (two-wire tester), as shown in Fig. 3-1. When the wire ends or "probes" are placed on the terminals of energized wires, the bulb will light. A charge of 230 volts will make the bulb glow brighter than one of 115 volts. The tester may be purchased at most hardware stores.

Fig. 3-1 Two-wire tester

Insert the probes into the receptacle. If the bulb *does not* light, there must be a break in the circuit from there to the fuse box, or a screw has become loose and fallen out either in the receptacle or on the fuse-box mounting plate. (We are assuming that the fuses or circuit breakers are all right, of course.)

If this bulb *does* light up, the next step is to determine if, with power at the receptacle, electricity is reaching the machine. One thing that might keep the power from reaching the machine is the use of extension cords to carry electricity to the appliances. This is not a particularly good idea, if for no other reason than that extension cords have caused many fires. Aside from that, every wire offers resistance to the passage of electricity, and the resistance of a thin wire, used in some extension cords, may be so great that the electric power will be reduced to a point where a motor will not run as it should. In this situation, it will not start readily and might overheat while running.

How do extension cords cause fires? Well, they are generally on the floor where they are stepped on, resulting in easily bruising or breaking the tiny strands of wire. The current arcing across broken strands gives off considerable heat, and, if the arcing is intense, the heat will burn off the insulation and proceed to char a wool carpet or anything else that is nearby and combustible.

To check the power input to the unit, it is necessary to remove the sheetmetal cabinet. This may be done by removing some screws. It is unlikely that you will have to use an Allen wrench or tiny screwdriver to get the knobs off; the cabinet generally has holes enabling the knobs to slide through. With the cover removed, and the controls and the control panel exposed, you will find the power cord leading to a terminal block. The cord should have a white wire on one of the terminal-block screws and a black wire on another one. If touching the test bulb probes to these two screws results in lighting the bulb, you can reassure yourself that the cord does not have a broken wire inside the insulation or a bad plug connection.

How a Unit Starts

At this point let us consider how the thermostat (see Fig. 3-2) starts the compressor. Remember that we are thinking of a simple, 115-volt air conditioner with only one thermostat; the heavy, complex units have more than one temperature-sensing device. In fact, at this stage we can even forget about the fan.

On the terminal block, the white wire indicates the "ground" side of the circuit, and the black wire indicates the "hot" side. The "hot" terminal will have a wire going to the thermostat. Remember, the thermostat is a device that can convert a change of temperature to mechanical movement. It has a bulb containing gas that clamps onto the evaporator, and this gas

Fig. 3-2 Typical thermostat

can flow through a tiny tube to a bellows, which is in a case or box mounted inside of the unit. (It is probably secured to the control panel.) When the surrounding air becomes warm, the gas in the bulb expands, flows through the tube, and exerts more pressure inside the bellows. The bellows, in turn, increases in size and pushes a mechanical linkage, which closes the contact points (see Fig. 3-3).

The thermostat is wired in series. That means that the current from the "hot" side of the input terminal block has to go through the contacts to complete its circuit, which is to the relay and on to the compressor motor. When the air in the room and the coil cool enough, the action is reversed—the gas in the bulb cools and the contacts open.

Fig. 3-3 Operation of a thermostat

Testing the Switch

If you want to know if the power is getting through to the relay, with its direct connection to the motor, test the thermostat. To test any "make-and-break" device, simply bypass it with a wire. In this case, if everything else is all right, you still can start the compressor with the contacts open.

If the motor starts, there is no question but that the switch is defective. If you want to quickly restore the unit to operation, pull the plug, place a jumper wire between the terminals of the switch (see Fig. 3-4) and plug in the unit again. That will make the unit run, and you can pull the plug again if the room gets too cool.

We have assumed that the terminal connections are of the screw type; if they are of the "knife-and-blade" type, you can pull off the blades and clamp them together with a spring-type clothes pin. This procedure does not provide a very good connection, however, and the unit should not be left running this way for any length of time.

Replacing the switch is a straightforward mechanical operation. There is something you must be careful about, however. The little tube cannot stand abusive treatment. If the tube becomes flattened, closing the internal passage, the tube is useless.

Fig. 3-4 Jumper wire on thermostat terminals

If an exact replacement cannot be obtained, a different switch capable of being physically mounted might work after a fashion, but, the various switches on the market have such different temperature ranges (on and off points) that, if a different switch were used, the thermostat's performance might be upset. If you tried interchanging switches between, say, a deep-freeze unit and an air conditioner, it probably would not work out very well. Of course, a certain amount of difference in the range can be compensated for by changing the dial setting. The defective switch might have the range printed on it, which can be used as a guide.

At this point, the current goes through the switch or bypasses it if you are using a jumper wire. A heavy-duty air conditioner might have a multitude of wires on its control panel, but what runs the motor is the circuit through the switch to the relay and the two wires from the relay to the starting and running windings of the compressor motor. Also, there will be a ground wire on the compressor motor that might lead back to the white wire terminal on the input terminal block.

Some people jump to the conclusion that, if the unit will not shut off, the relay contacts must have fused together. If you know the principle of operation, you will realize that, if the thermostat shuts off the supply· of current, the compressor will stop, regardless of the condition of the relay.

Consider the situation if the thermostat bellows or bulb develops a leak. When the gas is partly gone, a greater change of temperature will be required to expand (or contract) the remaining gas in order to move the contacts. Thus, the unit will be running longer than normal, and shutting off longer than normal. Continued leakage will exhaust so much gas that the contacts will not move at all; they will just remain open or shut—more likely open.

The Relay

The relay can energize the starting winding briefly because a motor draws a great deal more current when starting than it does after it reaches its rated speed. Any motor acts as a generator when it "gets going." This builds up a "counter flow" of current that opposes the flow of power from the fuse box. To be more exact, the rated-speed current is about 22 percent of the starting current. There are different types of relays, but in all of them the "high inrush" has the effect of moving a contact arm which closes the starting-winding circuit. When the "high inrush" drops off, the contacts open and disconnect the starting winding.

The circuit to the motor might contain one or more capacitors. They will be discussed later. There also will be a motor protector. Compressor motors always have a protector, although it will not necessarily be found on the fan motor. You can make a bypass test around a protector in the same manner as you did on the thermostat. These protectors are of the self-resetting type: if the motor heats up too much, the protector will open the circuit, and, when the motor cools off, it will close the circuit. The power flowing through the wire in the protector acts like an electric heater, and the heat causes mechanical action. Thus, in effect, it is a special purpose thermostat.

Common Electrical Terms

In the next chapter, we will learn about reasons why a compressor motor will sometimes only "buzz" and not turn over, or at least not fast enough so that the starting winding will disconnect. One of the reasons is poor voltage. But, before treating that subject, it would be well for us to have a clearer definition of "voltage" and related terms. We have been using "voltage," "current," and "power" rather loosely, and, for purposes of diagnosis of electrical troubles, we should know the meaning of these words more precisely.

Power is measured in watts and, actually, is a measure of how much heat is dissipated by anything receiving electrical energy. Current, measured in amperes, refers to the number of electrons flowing past a given point in one second. It is apparent that, if the current in an appliance becomes greater, more heat will be given off. The relationship between current, power, and voltage, as well as resistance, can be stated in several ways by formulas.

Voltage is electrical pressure, or force. The current has to have this "force" to overcome the resistance of the wires. If the force is weak, or "low," the apparatus involved, such as a motor, will not function as it should. Low voltage is a widespread problem. It might derive from inadequate power-distribution facilities or from inadequate wiring inside the house. Also, power plants are limited in the energy that they can send out through the wires. This limitation becomes quite troublesome at times of peak demand. Power companies are using "standby" equipment, such as gas turbines, to provide the extra generating power during the hour or two of the day when the current demand is the highest. If they did not, there would be a considerable voltage difference over the area being served.

Poor Power Supply

The problem derives from the fact that present-day usage of electrical equipment has quadrupled the requirements for power compared to the 1930's, when the requirements were not much different from what they had been for decades.

Consider this anology: If everyone on a street opened *all* their water faucets at the same time, the water pressure would fall and the main line under the street would not immediately be able to supply the increased water demand. In an electrical distribution circuit, when too much current is being "drawn off," the pressure or, as we say, the "voltage" drops. With a heating element, such as in a water heater, if the voltage is too low, it does not get as hot as it should. With a motor, on the other hand, low voltage might not permit it to start, and, if it does, it will run hotter then normal, which does not do it any good. In fact, you might say that a motor running on low voltage is destined to premature failure.

Air conditioners, in particular, are overloading the power circuits. A 230-volt air conditioner has two or three rather heavy motors. During a heat wave, they are running almost continuously. A washing machine, on the other hand, would be in use for a relatively short time.

The need for heavier lines increases continuously, but power companies are not able to string wires fast enough to compensate for the increased demand.

Aside from the limited funds power companies have available for modernization, many homes have the same small wiring that was put in them many years ago. And, it seems that they generally did not provide enough outlets, resulting in the use of extension cords which can cause a serious voltage drop. Even in newer homes, the occupants might have more appliances than the builder anticipated. No immediate, overall remedy is in sight.

Years ago manufacturers were able to help matters quite a bit by redesigning motors to provide a "good power factor," that is, to make better use of the power available. It has long been known that a motor may produce more torque than needed, thereby wasting electricity. In 1979, Exxon announced that they had perfected "electronic modulation," a means of instantly varying the characteristics of the current going into a motor, thereby providing some economy of operation. At present, the matter is the subject of a suit in a federal court to decide who will manufacture the equipment.

When an air conditioner with a low-voltage cut-out will not start, it could be that the cut-out has opened and not reset. If the input voltage is sufficient, the indicated repair is to replace the cut-out. Of course, you can bypass it, but then you will not be getting the protection you have paid for. If a timer is involved, bear in mind that every timer has to have its own little motor to make it advance. If the timer motor stops working, the dial will simply stay where it was. If the timer shuts off the power as it is supposed to, and then does not advance, the dial will have to be turned by hand.

The use of an extension cord with a unit introduces a fire hazard. It should be mentioned that the old wool carpets were not highly combustible; when they burned, the heat came mostly from the jute backing. The same cannot be said for synthetic wall-to-wall carpeting. A lot of it is now of a type that will start smoking as soon as the metal shield on a drop lamp has been left on it for several minutes, and the fumes it gives off are toxic. Commercial environments are required to use a more expensive type of carpet that has been treated to reduce the hazard. In rooms with a synthetic floor covering, it is advisable to reroute the cord so that it keeps clear of any combustible material.

4

Unit Hums, Will Not Start

Low voltage might be partly or entirely responsible for this condition. More exactly, either the compressor does not develop enough torque or something is preventing the compressor from turning as freely as it should. Or, it could be a combination of these conditions. Among the possible reasons for poor compressor torque are failure of the relay contacts to close, burning of the relay contacts, a defective starting or running capacitor, shorting of the running winding to the starting winding, shorting of either or both windings to the compressor-motor housing, or a binding fan motor that is drawing down the voltage.

As for the matter of interference with proper compressor movement, the trouble might be sticky or stuck bearings, or too much high-side pressure.

Testing the Voltage

If a low-voltage supply is suspected, it is very useful to have a voltage tester so that you can determine exactly what the voltage is and how much it drops when the compressor motor is energized. If you do not have one, you still might be able to learn whether or not the voltage is seriously low. In electrical work, there often are ways of arriving at conclusions by round-about means. In this case, you can move a TV set to the house wiring circuit under consideration and observe whether the picture is smaller than normal. If it is, there is no question but that the voltage is very poor.

If the voltage input to the terminal block is adequate, there might be a voltage drop beyond it caused by loose connections or frayed wiring (see

FRAYED STRAND

Fig. 4-1 Frayed copper strands reduce power input and drop voltage

CLEAN POINTS
RESULT IN GOOD
CONDUCTION

PITTED POINTS
RESULT IN POOR
CONDUCTION

(A)   (B)

Fig. 4-2   Pitted and burned contact points result in poor conduction of electricity

Fig. 4-1). Or the drop might be inside the relay. The relay has contact points similar to those in an automobile, and they also are subject to deterioration. When the contact points become pitted and burned, they cannot conduct the current necessary to start the motor (see Fig. 4-2).

Relay Failure

Why a particular relay fails is problematical. It might be that the bearings are sticky and the voltage is low with the result that, just after the points separate, the motor slows down, and the points close again. This situation might continue until the motor warms up (or cools off, as the case may be) and is able to maintain a constant speed of rotation. In this situation, the movable point might barely touch the other and move away. This erratic movement results in considerable arcing, and the heat given off damages the points. When the points become badly pitted and burned, the motor is not able to acquire enough speed to open the points; the motor will simply "hum" and, after a while, the motor protector will disconnect it from the power supply. This condition results from the windings drawing an excessive amount of power when the armature is in a stalled, or nearly stalled, condition.

When you hear the "hum," try gently tapping the relay. This might make the points come together a little better so that the unit can start. But, this action remedies matters temporarily only; the trouble will recur, and the indicated repair is replacement of the relay.

Faulty Capacitor

Another common cause for starting failure is a faulty capacitor. The early light-duty models were made with a capacitor only on the starting winding. The heavier models were provided with another one on the running winding as well. In the process of improving the power factor, the designers were able to improve the motor torque to a point where the capacitors could be eliminated on some models. Most units still have them, however.

Starting capacitors are usually electrolytic and are wired in series to the starting winding of the compressor motor. Running capacitors are oil filled and are connected between the leads for the starting and running windings. Typical capacitors are shown in Fig. 4-3.

Fig. 4-3 Typical capacitors (*Courtesy*, Frigidaire)

The capacitor is an interesting device; it has the ability to "store up" electrons and "give them back" the next time the current alternates. The effect of this is to improve the efficiency of the motor. The measure of a capacitor's storing ability is its "farad rating." A one-farad capacitor would be about the size of a ranch house. Practical capacitors are rated in millionths of a farad (microfarads).

Capacitors have some "weak points" that make them prime objects of suspicion when you have a will-not-start situation. The outside case, if it is plastic, can crack, and, if it is metal, can corrode. In either case, if the case opens, the electrolytic paste or oil can run out, thereby rendering it useless (see Fig. 4-4A). A capacitor cannot stand excessive voltage or too much heat, and, if an electrolytic capacitor was being energized every few seconds, it would be ruined before very long.

Capacitors deteriorate when not in use. If a unit has been shut off for several months and will not start up, it is very possible that the trouble will be traced to a capacitor. Carefully examine the case of the capacitor. If you find oil or paste oozing out, it is a positive indication of a defective capaci-

(A)   (B)

Fig. 4-4   (A) A cracked capacitor case permits fluid to leak out; (B) a bulging capacitor case indicates an internal short

tor. Or, if the case of an oil-filled capacitor is bulged, this indicates an internal short (see Fig. 4-4B).

A capacitor still might fail to perform properly even though it looks all right. Capacitors can be checked by certain instruments. If you do not have these instruments available, you still can check a capacitor by securing another of the same farad rating and connecting it as shown in Fig. 4-5. This procedure merely requires two pieces of wire, each having a clip on both ends. The illustration shows them connected in parallel, which is the

Fig. 4-5   Method of connecting capacitors in parallel

correct method. Of course, you can make a series connection so that the current will have to go through first one and then the other, but, if the suspected capacitor is "open," it would "block off" the flow of current through it, and you would not learn a thing.

To develop maximum torque, a motor has to have a capacitor of exactly the farad rating specified by the designers. A motor having a capacitor with a farad rating either more or less than specified will provide less torque, and it is quite possible that it will not start at all. For testing purposes, however, a capacitor having a farad rating somewhere in the general vicinity of the one being tested might work all right.

With some capacitors, there is the matter of ac polarity to bear in mind. With dc capacitors, the terminals have to be connected in a certain way with respect to whether they are marked positive or negative (plus or minus). When the ac capacitor was developed, it was designed so that it did not make any difference which way the two wires were connected. In recent years, however, a slightly different type has been specified for some installations. This type has to have the ground-side terminal connected as shown on the schematic for that particular unit. The terminal closet to the input from the switch is the "hot" terminal and the other one (identified by a spot of paint) is the ground, and, if you have the capacitor turned around so that the ground terminal is connected to the "hot" lead, the capacitor will not perform as it should.

Shorted Motor

Another reason for starting failure is a short in the compressor motor. You might have a short circuit between the starting and running windings (see Fig. 4-6) or between a winding and the motor housing (see Fig. 4-7). In either case, the electrical balance is upset, with resultant poor torque and heavy draw of current. This condition causes the protector to disconnect the motor before it reaches rated speed, or shortly thereafter. Later, we will show how the motor can be tested to provide a fairly good indication of what actually happened. If the tests indicate that the motor itself

Fig. 4-6   Shorted windings

Fig. 4-7   A short to the motor compressor shell causes arcing

is defective, little or nothing can be done about it. The motor-compressor will have to be replaced.

Likewise, in the matter of sticky bearings, if the condition is too far advanced, there is no remedy except replacement. Sticky bearings result from carbonization of the lubricating oil that has to be added to the Freon since it has little lubricating value of its own (some manufacturers used to add a form of alcohol to absorb the moisture and keep it from freezing). The presence of this anticarbonizing agent contributes to equipment failures.

Lubricant Breakdown

In refrigeration equipment such as air conditioners, the oil might serve as a lubricant indefinitely if it were not for the presence of moisture. Even microscopic particles of water cause a complex hydrocarbon reaction that results in the formation of a sticky compound that can adhere to the motor shaft. This "binding" of the shaft can progress to a point where the motor is unable to turn. The moisture, by itself, might not be too important if it were not for the heat which the water and oil are subjected to. A mechanism of this sort can get quite hot, and heat will accelerate these unwanted chemical reactions.

If the sealed unit is intact, we must assume that the moisture was in the air where the unit was manufactured. If the factories were in the Sahara Desert, it would simplify matters. However, they are located in relatively damp places, and the manufacturers take steps to dehumidify the air, and to bake the unit for a long time while holding it under a high vacuum. Even though advanced manufacturing techniques are practiced, there is always a possibility that a small amount of water vapor will remain in the system. If a unit has had six or eight years of use, it is likely that the bearings are at least a little sticky. It may be assumed, too, that the condition will get progressively worse.

High Head Pressure

If you encounter a hot motor that will not start, it does not necessarily mean that the motor has to be replaced. It might be that the heat has aggravated the sticking condition, and the shaft and compressor will turn when the motor cools. Actually, the condition is probably caused by high "head" pressure. The restrictor can pass only a limited amount of Freon, and, under certain circumstances, the high side, or "head" pressure can be quite high because liquid Freon is delivered to the restrictor faster than it can go through it on its way to the evaporator. If the unit stands idle for, say, 20 minutes, the refrigerant will have a chance to run through the restrictor and reduce the head pressure to relieve the stalled condition of the motor. Of course, the trouble might be caused by either hot bearings or high head pressure. In any event, the condition will usually remedy itself.

With these "will-not-start" situations, we have the classic pattern: the motor makes noises for a little while, perhaps five or ten seconds, until the "high inrush" or excessive draw of current causes the motor protector contacts to open. At that point, the motor stays "at rest" without electrical energy being delivered to it until such time as the protector points reset, which would be in a minute or two. Then the motor is energized again and the cycle is repeated. And that could go on until either the motor cooled off and started or you removed the plug.

Bypassing the Controls

It is apparent that the best way to determine whether the controls or the motor-compressor is responsible is to bypass all the controls and deliver power to the compressor motor directly. If the fan were wired to run with the compressor running, it would be taking power for its own use, which would, to some extent, reduce the voltage reaching the compressor motor. Aside from that, it would be time consuming to have to test all the wires and components on a complex unit individually to determine by elimination that the trouble is "apparently" in the compressor motor.

A three-wire test cord has been devised for this purpose (see Fig. 4-8). This tester has rubber-sheathed spring clips that can be attached to the three compressor-motor terminals. Such test cords are for sale in electrical supply stores. A professional test cord has a spring-loaded pushbutton that can be used to energize temporarily the wire leading to the starting-winding connection. If you wish, you can make your own (use heavy wire) with a toggle switch if you do not have a pushbutton switch handy. And, if you do make one, you will get along better with it if you fix the clips so that they can be attached at 70 or 80 degrees *away* from the perpendicular to the motor

Fig. 4-8 Three-wire test cord

housing. In the process of making units more compact, compressor motors have been installed in ways that do not make the terminals particularly easy to get at.

A factory-made cord should have the function of the wires indicated by color. By convention (national agreement) a white wire indicates a ground connection, a black wire indicates the "hot" lead, and red indicates the wire for the starting winding. This should hold true as far as the test cord is concerned; if the mechanism being tested has a few dozen wires, they might have many different colors.

Using the Test Cord

Now, what you must do with the test cord is to duplicate the action of the relay while providing an electrical supply direct from the receptacle. Here is how this is done: first, you must learn which terminal is which. You can determine this from the schematic, which should be pasted inside the cabinet somewhere. You *must* have the connections right. Do not guess. If the schematic is missing, defaced, or unclear, call the dealer. When you know which terminal is which, remove the wires from the terminals of the compressor, or, if it has a capacitor, from the capacitor input terminal. If you think you might not remember where they came from, make a drawing, and use bits of colored tape or tags to identify them. Attach the black-wire clip to the terminal for the running winding (*do not* have the test cord plugged in), and attach the red and white clips to the starting and ground terminals, respectively. When the test cord is plugged in, push the button to allow some of the black-wire current to flow in the red wire and energize the starting winding (see Fig. 4-9). Push the button for a few seconds (five

Fig. 4-9 Motor with three-wire test cord connected

seconds is the maximum). If the motor does not start in five seconds it is not going to, or at least not until it cools.

In properly made cord, the white-wire current will run through a fuse between the pushbutton and the plug. A stalled motor will blow a 15-ampere fuse. If the motor starts, and you interchange the black and red wires when you put them back, you then will have the relay disconnecting the running winding, leaving the starting winding to keep the motor going. That is not the way the motor was intended to run, and it will operate at higher than normal temperature. If you put the red wire on the ground terminal, on the other hand, when the relay disconnects, the motor will have an incomplete circuit and will stop.

Figure 4-10 shows a professional test cord with a knob that turns against spring pressure. The socket is for the insertion of a fuse. The two clips shown to the right are intended for a capacitor, but if one is not used, they are connected to each other. The small wire extending from the back side (below the power cord) is a ground cord that attaches to the compressor frame. The wire show at left has an attachment that goes on the modern terminal pin arrangement. The unattached jumper wire shown at the bottom attaches (as appropriate) to the terminals on the wire from the switch box.

Fig. 4-10 Compressor test cord

Refer back to Fig. 2-25 for various arrangements of the terminal pins. These wiring diagrams show things "as you look at them." In terms of the push-on connector for the three pins, they show what you see when you remove the connector and turn it over to expose the terminal clips. In other words, they are backwards. In terms of the compressor, they present the correct perspective. In any event, the compressor housing ought to have letters stamped near the pins to indicate their function: "C" for common, "S" for start, and "M" for main (or perhaps "R" for run).

In the interests of economy of operation, cooling equipment has compressor motors that are perhaps a little small with respect to the torque required to start the compressor. A larger motor could be used, but the monthly power consumption would be higher. Because the consumer insists that the equipment be as economical to operate as possible, we have the situation in which a unit might operate properly for some years; but, when the bearings become "sticky," there will not be sufficient motor torque to overcome the stickiness so that the motor will start and run reliably. It even could happen that the equipment will not start at the times of the day when the community power demand is high.

Can anything be done? Yes, very possibly. Where the voltage delivered to the unit is poor because of thin wires or poor connections, correcting these difficulties might keep the unit in operation for several more years. Heavier wires, preferably on a separate circuit from the fuse box with the terminal ends properly cleaned and the screws well tightened, might suffice.

If that does not do it, a special transformer might be used to step up the voltage. Whether or not you do this, however, should be determined by a person who is completely familiar with power distribution and motors. The "boost-and-buck" type of transformer is intended for this service. Such a transformer is designed so that the various windings can be interconnected to provide a great variety of output voltages. They can stand 24-hour use without overheating.

The "tapped-winding" transformer does not cost as much, but it has little flexibility. If the relationship of the input and output voltages happens to coincide with your needs, well and good, and if it does not, nothing can be done about it.

Sometimes the "low-voltage" difficulty stems from the fact that (to speak of the heavier equipment) two different voltages are being supplied. In some situations a 208-volt input is being supplied to equipment that should have 230 volts supplied to it. The reason that tapped-winding transformers were placed on the market was to convert 208 volts to 230. They can do so, but, if the input is not even 208, or if it fluctuates considerably, the "boost-and-buck" type is more useful.

It should be noted that some 277-volt air conditioners are on the market. The supply wires for this voltage have to run to a 440-volt transformer. Before buying one of these units, it would be advisable to check with your power company to learn what the expense will be for installation of that voltage service.

Incidentally, it is remotely possible that the reason for the low voltage is that the power company has its distribution taps improperly connected. In that (unlikely) event, they would remedy the situation.

Sometimes fans will not start. They, too, have capacitors. Here again, an exact replacement is advisable. In the oil-filled type, there is a voltage rating as well as a farad rating. If you get a replacement made by the same company, it will have the same physical dimensions and you will not have to alter the mounting bracket to use it.

5

Unit Runs, Will Not Cool

Low Voltage

Here again, low voltage might enter the picture. If the unit is started when the power demand of the area is fairly low, it will operate at proper voltage. As hours pass, the power demand might rise, causing the unit to run on a lower voltage. For example, if it is started at 9 A.M., that would be a time of the day associated with low household-power consumption. If it is still running at 11:30 A.M., a voltage drop would be encountered because people would be preparing meals and their ranges, electric fryers, and such would be in use. The low voltage condition, if serious, would slow down the motors.

As was mentioned, the unit cannot cool if the air is not moving normally. This statement applies to the movement of air through the evaporator coils and to the flow of air carrying-away the heat of the compressor and condenser. If either fan stops, the room will warm up.

What we will concern ourselves with directly at this point, however, are reasons for the temperature of the evaporator coils not being as low as it should. This situation might stem from a restriction (obstruction) in the system, too much or too little refrigerant, air in the system, or a dirty condenser. Needless to say, having the hot sun shining directly on the condenser does not help matters either; the ability of a condenser to give off heat is dependent on the difference in temperature between it and the air, and the heat of the sun reduces the differential.

In addition to the foregoing, you might have a condition in which a malfunction of the compressor (stuck or broken part) results in the pressure equalizing on both sides of the system. In other words the pump was not pumping.

Restrictions

Suppose it was a restriction; this situation is similar to the condition described in Chapter 4 in which "sticky" bearings resulted from carbonization of the oil. Actually, when the oil carbonizes, minute particles are circulated with the Freon, until they finally adhere to some part. In this case, they adhere to the walls of the restrictor.

What is this restrictor? Recall that we said earlier that the liquid Freon has to be held back, or restricted, so that the drops have a chance to vaporize? In small units, this function is accomplished by having the refrigerant flow through a length of spiral tubing so small that its internal diameter is measured in thousandths of an inch (about 0.035 inch). When the refrigerant flows through this tube, its flow is "restricted" by the friction developed between the fluid and the walls of the tube. Thus, the size and length of the tube used permits the right amount of Freon to flow through it, unless something reduces the size of the internal passage. When an impurity gets caught in the restrictor, microscopic particles of carbonized oil build up on the walls and around the impurity until the Freon flow is stopped (see Fig. 5-1). You might have a restriction at other points in the system, but the restrictor is the most likely place for it. A serious restriction usually calls for the replacement of the unit.

Fig. 5-1 Clean and carbonized restrictors

On the other hand, if the passage is clear but the system has too much refrigerant flowing through, the performance will be poor also. The evaporator will fill up, or "flood," and the Freon will be vaporizing in the so-called "suction" tube, which is the larger of the two tubes between the evaporator and the compressor. In this situation, you can detect the condition by sight and touch. If the weather conditions are right, the suction tube will actually frost up, or it will at least be cold and collect moisture.

Too Much Refrigerant

How could the system have too much refrigerant? It could have been accidentally overfilled when it was being repaired, and it's not inconceivable that it was overfilled at the factory. One some models, the refrigerant can be drained from the system by loosening a screw on the compressor "charging stud" (see Fig. 5-2). This condition can exist for quite sometime before it requires investigation, that is, at times when the unit runs for only a few

WRAP AROUND CONDENSER DISCHARGE LINE COMPRESSOR CHARGING SCREW

DEFROST TIMER DRIER-FILTER CAP LINE SUCTION LINE

Fig. 5-2 Refrigeration system of household refrigerator (*Courtesy,* Frigidaire)

minutes and shuts off, only a moderate amount of refrigerant is pumped into the evaporator, and it does not flood. By the same token, it follows that an air conditioner could serve well enough during several weeks of mild weather and then give trouble during a hot spell.

Putting in refrigerant is called "charging" the unit, and, when it leaks out or is drained, the unit is said to be "discharged." On a partly discharged unit, you can see an incomplete frost pattern on the evaporator, or you can feel the coils on the turns not covered by fins, to determine which are cold. As the leakage progresses, two-thirds of the evaporator will be cold, then one-third, and, finally, it will not be cold at all. If all the turns are warm, you can not tell without instruments whether the refrigerant is discharged or if there is a restriction.

There is another procedure that might be tried. Applying a little heat to the restrictor might enlarge the tubing to dislodge the restriction. The spiraled restrictor tubing is contained in a bulb in the small tube from the compressor (see Fig. 5-1) at the point where the compressor is connected to the evaporator. You must be exceedingly careful in applying the heat, however, because the restrictor has soldered connections that can easily melt. This method sometimes works, but if the "slug" of dirty oil is permitted to circulate around the system, it will, sooner or later, cause another restriction. You might ask, "Why can't the dirty oil be drained and fresh refrigerant replaced?" This is done on heavy commercial equipment, but, on household units, it is usually not feasible.

Halogen Leak Tester

If the unit has discharged, it is obvious that there must be a leak somewhere. Any of the soldered joints can leak, or the charging stud (sometimes called a valve), or the terminals. The tubing itself might develop a "pin-hole" leak, usually on a sharp bend, where the deformation resulting from the bending has weakened the wall.

When refrigerant leaks, the lubricant usually leaks with it. If you can find the spot where oil seems to be running out, this can lead directly to the leak; at least this is the case in a refrigerator. On an air conditioner having a fan motor that requires periodic oiling, you might not know for certain whether or not a spot of oil resulted from a leak or from oiling the fan. With the fan, there is so much moisture (from rain water), and turbulence of the air, that oil is easily distributed.

Nevertheless, if a leak appears, what is the testing procedure that can be used to discover its source? Probably the most common testing tool in use is the flame-type leak tester. This device draws air, needed for combustion, through a rubber tube. Placing the end of the tube at a leakage point, you will observe that a small leak changes the flame color to lavender, and a large leak changes it to green.

Domestic fuel gas and various other fumes can cause slight variations in the flame color. Some servicemen think they can distinguish one impurity in the air from another by its effect on the flame, but the variations are minute enough to be difficult to detect even by an expert. Before testing an air conditioner, you must disconnect the fan because the movement of air, besides making the flame flutter, will carry the refrigerant fumes away from the point of leakage.

Figure 5-3 shows how a flame tester is used. The one pictured is alcohol-fired; a propane-fired type is also available.

Another testing tool is the electronic leak gun (Fig. 5-4). Robinair's explanation of its operation is as follows:

"It has been discovered that low intensity voltage pulses can be interrupted in an atmosphere having a different dielectric breakdown potential than that of the surrounding atmosphere. All-halogen based refrigerants have this characteristic and make it possible to adapt this new principle to leak detection.

Fig. 5-3 Testing for leaks with a halogen tester

CUT-AWAY VIEW

SENSOR PROBE

REFERENCE LEAK

ELECTRODE CAGE

AUDIO SIGNAL

OSCILLATOR CIRCUITRY

TRANSISTORIZED CIRCUITRY

"D" CELL BATTERIES

TRIGGER KNOB
"ON-OFF" SWITCH

Fig. 5-4 Electronic leak gun (*Courtesy*, Robinair)

"Under normal operating conditions the electric discharge pulses are greater than the dielectric breakdown potential of the atmosphere being tested, and no leak signal is indicated. However, when minute traces of halogen gas enter the electrode cage, the dielectric breakdown potential of the atmosphere is changed. The voltage pulses cease and the Leak Gun emits a piercing audible signal.

"Because of the simplicity of the Dielectric Differential Principle, the Leak Gun requires only one combination switch to ready it for use. The trigger knob doubles as an off-on switch and sensitivity control. On turning the trigger knob On, a piercing squeal is heard. Continue turning knob in same direction until tone stops. At this setting you are ready to probe for leaks. Built-in reference leak permits actual check. While most leaks will be found with the probe positioned on the gun, it can be removed and extended up to five feet on its flexible lead for locating leaks in hard-to-get areas."

In 1979 leak guns of this sort were selling for about $120—without attachments.

A shortage of refrigerant will change normal operating pressures. Whatever the pressures may be, they vary according to temperature. When the surrounding air heats up, they increase; when it cools off, they go down. With too little refrigerant, the compressor will not receive vapor fast enough to satisfy its capacity. It will keep running at normal speed with the result that the low-side pressure will drop to a *very* low level. Pressures run higher with R-22—a low-side pressure of 40 pounds, perhaps. With R-12, the pressure will probably be in the 4- to 10-pound range. Should there be a leak, the compressor will try so hard to maintain the correct amount of liquid on the high side that the low-side pressure will keep dropping until the compressor is finally running in a vacuum. This is a bad situation because the air drawn in as the result of a low-side leak may contain moisture that will carbonize the oil. The pump will labor trying to push the air into the evaporator, and the condenser will become very hot.

If the low side operates in a vacuum, you will have to add refrigerant to locate a leak there. Enough internal pressure has to be created at the point of leakage for the leak to be detectable. Recharging will be described later.

Malfunctioning Compressor Valve

Another reason for little or no cooling is the malfunction of whatever serves as a valve in the compressor. The manufacturers have various notions about how compressors should be built, but they all have to have an arrangement to prevent the vapor from flowing back to the low side while it is being compressed.

One model uses a solid, off-center wheel inside of a cylinder (see Fig. 5-5). The wheel has a "divider block" bearing against it. This block slides in

Fig. 5-5   Divider-block compressor

a groove and has a spring that holds it against the wheel. The eccentric mounting causes the wheel to move toward and away from the cylinder wall at the point where the divider block is located. The movement of the wheel alternately covers and uncovers two openings or "ports." One of these ports leads to the low side, the other to the high side. When the low-side port is uncovered, the heat-laden vapor from the evaporator flows into the space between the eccentric wheel and the cylinder wall. Then, the rotation of the eccentric wheel pushes the vapor against the divider block and squeezes it out through the high-side discharge port, thereby condensing it.

If the divider block cannot move freely in the groove because oil has adhered to it, it will be pushed to the end of its travel and stay there. You then would have a situation in which the piston moves air in and out through the valve openings, but no compression is taking place. This results in equalization of pressure on both sides of the compressor. A situation with similar symptoms can occur on a heat-pump model when the reversing valve leaks.

Dirty Condenser

If the condenser becomes covered with dirt (see Fig. 5-6), that, too, can nullify the cooling action. The compressor delivers its compressed vapor to a condenser, and, if the condenser stops working, the pressure will build up even though the refrigerant will bleed through the restrictor.

When a coil becomes covered with dirt, it is often said to be "loaded up," and this confuses people who do electrical work because they automatically think of the term in an electrical sense. Actually, in a refrigeration unit, you can "load up" a coil in three different ways: a coil of

Fig. 5-6 Dirt on a condenser prevents it from cooling Freon refrigerant

wire, such as a solenoid, can be loaded electrically; a condenser or evaporator coil can be, on the outside, physically loaded with dirt; and a condenser or evaporator coil can be loaded up on the inside with liquid Freon (implying that it was "flooded," with little vaporization taking place).

With a split-type air conditioning system, it sometimes happens that the screen of the outside part becomes covered with an accumulation of dead leaves or scraps of paper. Upon their removal, the unit should revert to normal operation.

6

Dial and Cycle Defrost Refrigerators

Air Circulation

The early refrigerators with a "freeze box" at the top, cooled the food entirely by convection. That is, the air that touched the freezer would chill, become more dense, and settle to the bottom. This dense cold air would displace the air at the bottom, pushing it along the sides forming a continuous circulation pattern (see Fig. 6-1).

This considerable movement of air, combined with the frost gathering tendency of the freezer, tended to dry out the food. To combat this tendency, glass-covered hydrator trays were put in the bottom to conserve food moisture in such things as green leafy vegetables.

Then it was decided to put the evaporator coils in the walls so that the walls of the food compartment could cool by radiation. Many ideas were

Fig. 6-1 Air circulation pattern in old-style refrigerator with center freeze box

Fig. 6-2 Air-circulation pattern in refrigerator with "across-the-top" freezer

worked out with respect to coil placement and air circulation; the outcome was that the "across-the-top" freezer was introduced.

Here, again, we have convection, but, in this case, the chilled air flows down outside the food space. Actually, it drops mostly along the inside of the door (see Fig. 6-2). Although warmer, moisture-laden air will have to make its way up to the top and around the freezer, much of the warm air goes up the center and hits the plastic tray, which is not cold enough to accumulate frost.

The designers anticipated certain undesirable effects from the convection, but the varying climatic and usage conditions make it virtually impossible to design a unit that would work perfectly in every situation. In certain instances, the manufacturers have provided a baffle to compensate for unusual operating conditions, but placing the unit in a cold room or on a sun porch will upset its operation. If the "cold control" is set to maintain a food-compartment temperature of 40°F, and the outside air is 40°F, the unit will not operate and the freezer will attain the same temperature as the food compartment.

Loose Door Seal

With either type of freezer, a leaking door seal can cause longer-than-normal running time and excessive frosting. A leak can be due to rotting of the seal (probably caused by fruit juices), flattening due to normal deterioration, a sprung door, or a loose latch. You can check tightness of the seal by taking a dollar bill and closing the door against it; you should feel some tension when pulling the bill out.

(A) MAGNETIC TYPES

(B) COMPRESSION TYPES

Fig. 6-3 Typical door seals or gaskets

If there is generalized looseness caused by flattening or wearing of the door latch and striker (the stationary part), it can be corrected by loosening a few screws, moving the striker and retightening the screws. If the door is sprung, you might be able to put it back in shape by pulling on the top while bracing the bottom with your knee. Some of the bigger doors have tie rods inside them that can be adjusted. It is quite a lot of trouble to get at them, however; you would have to remove the door-liner panel.

If the seal has to be replaced, it can be removed by taking out some screws. Probably several hundred varieties of seals have been marketed (see Fig. 6-3). There are several methods of attachment, but, usually, the screws are underneath the semicircular part. The larger cities have supply houses that sell the more common seals in 14-ft lengths.

Because seals are made of flexible material, they do not provide even pressure all around the door. This does not matter with the older models because you can adjust the striker so that the door closes tightly. With the newer models that have magnetic door closure, a "thin" spot will cause a leakage of air that, around the freezer at least, will produce an accumulation of frost. The remedy for that is to insert shims behind the seal, building it up so that it provides even pressure. It does not matter what is used—even cardboard shims can be used, provided it is thin cardboard. Measure the distance between screw heads, then cut the shim(s) so that they can be placed between the screws or on both sides of a screw.

Erratic Lights

Cabinet lights sometimes flicker, and the trouble may simply be that the bulb is not screwed in far enough to make proper contact. Some bulbs are recessed and covered with a glass or plastic shield. If no screw heads are showing, the shield is held in with spring clips and perhaps a piece of rubber. It can be removed by pressing on it and sliding it one way or another. When the right direction to slide it is found, it will come free.

Figure 6-4 shows a refrigerator light socket in which the rubber sheath bears against the glass on the replacement bulb in a way that makes it appear that the bulb has been completely screwed in when in fact it has not been.

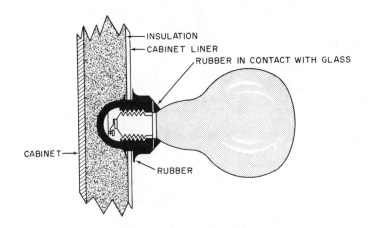

Fig. 6-4   Refrigerator light socket

Faulty Latch

How would you open the door if the handle came off? At the worst you could remove the hinges, which should be easy. Some hinge butts may be sprayed with a coat of paint, but usually they have a sheet-metal cover over the screw heads. These covers fit the butt castings perfectly, and stay on because, on the top and bottom, there are little indentations that fit into depressions in the casting beneath. There should be a turned-up edge so that a knife or small screwdriver can be inserted for removal; they snap back into place.

The door closure is in two parts. One part, the latch, "latches" onto the striker which is secured to the cabinet frame. Actually, the terms "latch" and "striker" are used interchangeably, and, therefore, you should exercise care when ordering one of these parts so that you are not misunderstood and receive the wrong part.

When you pull the handle, the base of the handle presses against a plunger that, through a linkage, moves the latch up or down or sideways to disengage the striker (see Fig. 6-5). Normal wear will make it necessary to pull the handle back more and more before the slack is taken up and the

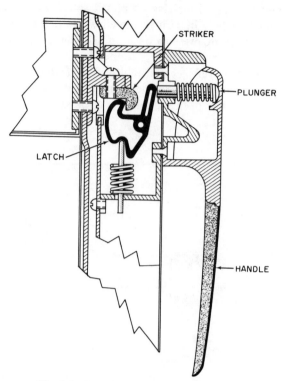

Fig. 6-5 Typical door closure assembly

desired action of the latch takes place. Washers or spacers often can be installed to compensate for the wear. If the handle breaks, the easiest way to open the door is to remove the handle mounting bracket and manually push the plunger. To get at the bracket, pry off the sheet-metal covering of the handle, thereby exposing the screws of the bracket.

If the linkage is broken, this idea will not work, because the hole in the door is too small to reach through with tools.

With an "up-and-down" closure, an enlargement on the end of the latch runs between two nylon rollers (see Fig. 6-6). One roller is fixed and the other moves against a spring. To open the door, merely slide a table knife between the cabinet and the rubber seal, and exert pressure as required (usually downward), while slightly pulling the door.

It is possible for the striker assembly to jam. Before sawing off the latch with a bayonet-type hacksaw, try removing the hinges and reaching in from the hinge side to smash the plastic molding covering the screws holding the striker. Replacing the plastic molding is simple compared with replacing the latch. A door liner has numerous fasteners and the linkage has tape and insulation around it. When the liner is removed, the insulation has

Fig. 6-6   Typical "up-and-down" door closure assembly

a tendency to move, and you will have to keep the insulation in place while you adjust the liner for fastening.

If a "sideways" closure linkage breaks, you need only remove the hinge screws on the cabinet part; the other stays fastened to the door. Once this has been done, it is a simple matter to move the door sideways to disengage the latch from the striker.

Of course, the machine might have to be moved away from the wall, and we should consider how that can best be done. It is fairly easy to move a refrigerator around the room if you slide it on something. Linoleum installers use a felt blanket, but any scatter rug will do. You will have to push at a low point to avoid tipping it over. The simplest procedure is to throw a loop of clothesline over the refrigerator and pull on the bottom.

Refrigerator Relay

Relating the refrigerator mechanism to the material in the foregoing chapters, and following the same general order of items, we can say that the refrigerator has the same series-wired switch, relay, and three-wire compressor. The relay has the leads for the motor protector, which may be built within or mounted beside it. There would be power-line connections and either three wires leading to the compressor or three clips for terminal pins. There would be two wires for the cabinet light switch and two for the thermoswitch, unless there is a common ground, in which case there would be three. In one wiring system, the light wires connect in the relay and in another they bypass it. Here again, you should find a schematic that will tell you what the wires are for and where the connections are made. As with any household equipment, you can duplicate the relay action and start the motor with a three-wire test cord.

In Fig. 6-7 we have one type of relay. It is a "thermally operated starting and overload relay with complete compensation for room temperature." The engineering of such a device is a bit involved, but, simply stated,

Fig. 6-7 Thermal relay assembly

the principle of operation is this: the current from the switch enters through terminal "L" and goes to the running winding through terminal "M." It also goes to the starting winding through terminal "S." The passage of current through the bimetal starting arm causes it to swing to the right, thereby separating the contact points. The heater element supplies enough heat to keep the starting arm in the "contacts-open" position. On the right hand side, we have the bimetal motor-overload arm. This arm is made so that it will not bend as readily as the starting arm, but, if the motor draws an excessive amount of current, it too will bend, thereby opening the contacts.

If the current draw is heavy enough to open the overload contacts, it will heat the reset bimetal enough to bend it, which has the effect of shutting off the heater. At the same time, the tip of the reset bimetal will move to the left and touch the starting arm so that it will more readily return to the "closed-contact" position.

The overload arm is made so that it will not cool as quickly as the starting arm, so that the starting contacts close first. With both sets of contacts closed, the relay is ready to start the motor again.

The bimetal compensating arms will bend and vary the position of the toggle springs so that the operation of the relay will not be affected by changes in room temperature. Without the toggle springs, it is obvious that a device of this sort would be unreliable if the room temperature was very low or very high.

Although the relay might have nine or more wires, it can be replaced, without confusion, if you change the wires on one terminal at a time. Some of the manufacturers have used several different relays on a particular size of compressor, but some standardization is provided by requiring only one relay as a replacement for all of them. In this event, the connections might not be exactly the same, and, hopefully, a drawing will be furnished with the new relay to show what wiring changes to make. Figure 6-8 shows another type.

STATIONARY CONTACT

GUIDE PIN

BRIDGING CONTACT

STATIONARY CONTACT

SOLENOID COIL

PIN CONNECTORS

SPRING

ARMATURE

Fig. 6-8   Current-type relay

Operating Cycle

A partly discharged cold control bellows will produce alternate freezing and defrosting. The defrosting will take place throughout the freezer instead of just parts of it as happens when there is a shortage of refrigerant. With the dial defrost type of unit, a fairly short cycle is normal. (Of course, switch cycles are affected by climate and usage. The normal cycle with an ambient air temperature of 70°F depends on the characteristics of the switch involved.) A run of four minutes and shut-off for six might be typical. With the cycle defrost type of unit, on the other hand, a long cycle is normal, but it should never shut off so long that frozen food thaws. More will be said about cycles later.

Replacing a Cold Control

The cold control is behind the dial and can be reached by removing the ornamental fixture covering it. The capillary tube will extend through the insulation space and lead into the food compartment and connect to the freezer. The tube might extend for quite a distance between the insulation and the cabinet liner; it might even go around a corner. When you pull the tube out, you would do well to tie a heavy string to the other end so as to be able to pull it back through the insulation.

What was said before (of air conditioners) about low voltage and starting troubles also applies to refrigerators. Using a three-wire test cord, a 1/7- or 7/32-hp compressor should start without blowing a 7½-ampere fuse.

The refrigerator will not have a running capacitor, but you might find a starting capacitor. You can learn something about the condition of the capacitor by removing the terminal wires, briefly energizing it with the test cord, and then holding a jumper wire on one terminal and touching it to the other. This procedure should result in an arc. When testing small capacitors in this fashion, the normal arc will be so small that you can just hear the "pop" above the surrounding noise level. In other words, the test would not be very conclusive. On a relatively large capacitor, however, if it popped like a firecracker, you could rest assured that the capacitor was in good condition.

If a unit does not start well, you can insert a starting capacitor. The dealer can advise you as to what farad rating should be used, but, if you do not have the factory recommendation for refrigerator motors of 1/7 to 1/8 hp, starting capacitors in the 105-145 microfarad range have been used satisfactorily. Even if the motor will start, it might be well to write the factory and find out what size is recommended. You still might not be getting maximum torque, and, as a consequence, hard starting would be encountered when, for instance, the voltage drops off a little.

In a matter of poor cooling, a possible cause is that normal wear in the mechanical action of the cold control has had the effect of changing the temperature range of the switch. This is nothing to be concerned about; just turn the dial to a colder setting. If the refrigerator runs all the time but will not cool, however, it is worthwhile to consider whether some household items or newspapers have fallen down in back from the top of the refrigerator. Newspapers in particular can drop down and not be noticed, inhibiting air flow to the mechanism, which will result in an increased cabinet temperature. Some refrigerator units on the market have fans for condensers. Here again, the fan might have stopped, or the condenser might be covered with dust.

It is normal for the warm liquid refrigerant to defrost the evaporator in a very small area where the restrictor is connected. The size of the spot depends on the model. If the unit starts to discharge, however, defrosting is noticed at the point where the heat-laden vapor is drawn into the suction line, that is, where the larger tube is connected. The refrigerant can absorb only so much heat, and when it is fully heat laden, no cooling of the tubes can take place in the area through which the hot refrigerant must flow on its return to the compressor.

Overcharging

The volume of refrigerant governs what portion of the freezer is defrosted. The "defrost area" starts where the vapor leaves the evaporator and grows progressively larger as the Freon leaks out. After a while, the defrosted area extends to the restrictor. The discharging takes place over a period of weeks, unless you have a terrific leak, in which case you can hear the hissing, if the general noise level is not quite high.

An overcharge will show up by water dripping on the floor caused by the melting of the frost that forms on the suction line. You will know if the refrigerator is overcharged because the suction line is normally warm, not cold. "Normal" implies starting and stopping (cycling) without the unit doing a great deal more cooling than it would if its contents were already cold and with the door not being opened many times. When first installed and started up, when starting after defrosting, or after having considerable hot food put into it, the suction line can get cold without indicating an overcharge. The condition will remedy itself when the inside temperature drops and the unit runs on a short cycle again.

If the symptoms indicate that there might be a restriction, the indicated procedure is to try to locate it. (A restriction does not have to be in the restrictor.) If the restriction is not completely blocking the flow of refrigerant, there should be some frosting in the general area of the restriction. But, sometimes, the restriction just cannot be located, resulting in replacement of some or all of the sealed unit parts.

Adding Refrigerant

To put in refrigerant, you must create an opening in the sealed system by means of a charging stud, if there is one, or with a clamp-on attachment that will puncture the tubing. By all means, charge on the low pressure side of the system. To use a charging stud you have to have a charging tool, the principle of which is shown in Fig. 6-9. The tool attaches to the charging stud at one end; the other end has a packing with a valve stem passing through it. The internal end has a means of turning the screw within the stud. With General Electric and Hotpoint compressors, the screw has a hole that a spline on the tool will fit into. In this case, pushing on the stem will push the spline into the hole; it can then be turned to the left, thereby opening a slot through which the refrigerant can flow in or out of the compressor.

Fig. 6-9   Charging stud and charging tool

As for the clamp-on fittings, several types are available from appliance supply stores. There are two general groups: One kind comes in two sections on opposite sides of the tubing that are drawn together with a number of screws. The other is the "yoke" kind that slides over the tube (see Fig. 8-3). In this last group, a nut on the side opposite the handle is drawn up to hold the fitting securely in place. A popular variation of this features a "depressor valve" instead of a screw stem. The depressor valve, also called a "bicycle valve," is very similar to the tire valve on a car.

The original design of the other group, where the parts are drawn together with allen screws, called for yet another pointed allen screw in the stem. This was screwed in initially to puncture the tubing, then unscrewed to make an opening, and finally screwed in again to close that opening. Obviously, there is a possibility of a leak with this design. The group, however, is also available with a depressor fitting. The advantage of a depressor is that the more pressure there is, the tighter the seal. A depressor hole is larger, moreover, thereby allowing a more rapid passage of the gas. At least, that is the case that is made for the allen screw types. To get a good seal, the screws have to be tightened evenly, of course, and it is therefore desirable to find a straight place on the suction tubing to put on the fitting. When tubing is bent, it has a tendency to become oval shaped, and the deformation cannot be gotten out of it.

Standard flare fittings are ¼-inch, and a special hose is needed for a depressor fitting. R-12 and R-22 as well as the hose can be bought in automotive stores.

The refrigerant comes in cans of about 1- or 2-lb capacity. The same supplier will provide some sort of screw-on or clamp-on fitting to open the can. With everything ready, what then? You have to get the moisture-laden air out of the hose. To do so, leave the nut on the charging fitting loose and let in gas to expel the air. When you think it is out of the hose, tighten the nut so that the charging fitting can be opened. The can has to be right side up for vapor to come out instead of liquid. *Caution*: Remember that *liquid refrigerant can blind you.* When the evaporator starts to get cold, you might as well close the valve on the can.

Sometimes you can pinpoint a leak by the oil that has run out. If not, you have to search for it. The search might lead to the area behind the plastic molding where the tube goes from the bottom to the top. Some moldings are attached with screws; if not, you might have a problem on your hands. Without the use of screws the molding is fastened by spring clips or flexible "dog ears." The trouble arises from the fact that you do not know which of these two methods has been used. Begin by using a table knife to break the seal between the molding and the cabinet on one side, and the cabinet liner on the other. Then exert pressure to slide the molding toward the food compartment. If it is held with spring clips, they can be bent somewhat by pulling on the edge of the molding (see Fig. 6-10).

Fig. 6-10 Typical plastic molding placement

Soldering

Charging stud and terminal leaks are repaired mechanically by tightening. Elsewhere, it is necessary to fill the hole with solder. Refrigerator soldering is difficult because it frequently involves soldering either the bottom side of a horizontal tube or soldering on a vertical surface.

Leaks are often found in the evaporator. The evaporator is made of aluminum, which easily can be warped by heat. Furthermore, ordinary solder does not bond with aluminum particularly well. A special low-melting-point aluminum-soldering kit is marketed by the Marquette Company, among others, and is available in automotive supply stores. Aluminum solder comes in different temperature ranges. You need the kind that will melt at 400 °F or less.

To solder anything, it must have a clean surface. In a freezer, you will have some oil to contend with, in addition to a film of food vapor. The area surrounding the leak can be cleaned with carbon tetrachloride; but be careful: this solvent is highly toxic.

A freezer presents a large metal area that continuously draws away the heat applied to the point you are soldering. The metal should be hot enough to melt the solder upon application, but this is difficult to attain because, in using any amount of heat, you run the risk of warping or even melting the sheet metal. In addition, the heat carbonizes the oil inside and might cause a chemical breakdown of the Freon itself. Also, you have to use flux, and if some of it gets through the hole, it will start a reaction of its own.

One procedure uses both a propane torch and a "pencil" type soldering iron. (You must clean any corrosion off the "pencil" point before you use it.) The propane torch is used to heat the leak area to decrease the heat dissipated by the metal. After heating the point of leakage, apply the flux. (If the flux changes color, the metal is too hot; allow the metal to cool.) Then place the end of the solder strip at the hole and use the pencil point to melt a drop of it, pushing it into the hole. A pencil-point soldering iron works quite nicely for this because the ordinary freezer leak is only a pin hole.

Sealing with Cement

Another method of sealing is to use epoxolytic cement. A tube of it can be bought at a hardware store and used according to the directions. Epoxolytic cement is simple to use and has been used successfully, but whether epoxolytic sealing will supplant soldering for evaporator leaks is a question that stands unresolved so far.

Because the freezer is on the low, or suction, side of the system, it follows that some air might have been drawn in through the leak. In theory, you should use a vacuum pump to remove all the air and Freon (called evacuating) and recharge the unit every time a leak is discovered. The evacuating procedure will be described in Chapter 8.

When determining whether it is worthwhile to search for a leak, it is helpful to know the running time of the unit. When the refrigerant charge decreases, the running time increases. In other words, if it takes one hour of running time to complete the cycle with a certain amount of Freon, it will take two hours to attain the same cooling effect with half as much. Maybe it does not work exactly in that manner, but that is the general idea. However, the power consumption is reduced with a shortage of refrigerant.

Checking the Cycle

Modern refrigerators run so quietly that most of the time the housewife simply would not know how long it was running. We know that a sticky-bearing-and-poor-relay situation might prevent it from operating on a normal cycle, even if it appeared all right when listening to it. And, in terms of the cooling that the unit is required to do, it might be that the cooling capacity is exceeded because of high room temperature, a leaking seal, or a dirty condenser.

If, in the analysis, the running time of the refrigerator is determined, you can use that as a stepping stone to further deductions. One way to do this is to take an electric clock and wire it to run when the compressor does. The clock can be attached to the white terminal in the relay and to the running winding terminal.

If the unit runs most of the time and does not cool, it may be presumed that, if the unit starts and stops occasionally, the switch and relay are all right. If the condenser is very hot, there is probably air in the system.

If the unit runs all the time and does not cool, which could very well happen in a hot climate or in summer, the answer might be simply that an excessive number of beverage bottles or cans were being chilled. Chilling warm liquid will make the unit run more than anything else.

Volt-Wattmeter Readings

Although the foregoing procedures may be used to arrive at a deduction, they do not provide a means of quickly getting a direct indication of what is wrong. A volt-wattmeter can be used for that, but even this is not 100% foolproof. You can have a combination of things wrong that will throw the readings off. For instance, you can have a normal wattage reading with sticky bearings and loss of refrigerant.

With the volt-wattmeter shown in Fig. 6-11, one can read watts on the left and volts on the right. This model has three wattage ranges and two for volts, while some have only two wattage ranges.

Power can be computed by taking the square of the voltage divided by

Fig. 6-11   A volt-wattmeter (*Courtesy*, Robinair)

the resistance, among other ways. Rewriting this expression using Ohm's law (E = IR), will reveal that voltage has some connection with the wattage reading. The voltmeter will show how low the voltage drops at the time of high inrush when the unit is starting.

When using a volt-wattmeter, set it on high range for starting and switch to low range after the motor has speeded up. Several things can be determined with this device. With a normal start, the needle will "over-shoot," and travel beyond the reading for the starting and running windings combined, but it will quickly drop back to where it should be, and, when the starting winding disconnects, it will drop to the running-winding reading.

With a normal start, the running winding will energize quickly, and the needle will "dwell" only momentarily on the starting reading. Thus, the meter will tell you the time required to energize the starting winding. In other words, it tells what the relay is doing.

With a cold motor, the reading will drop somewhat as the motor warms up. You will not learn very much when the unit starts. You have to wait until the refrigerant circulates to establish a normal relationship between the high- and low-side pressures. How long that will take depends on circumstances, but it might be nearer 20 minutes than 5.

Because the pressure of the refrigerant is related to air temperature and the capability of the condenser to give off heat, and because consumption of power by the motor is related to the voltage received, you will get a different "normal" reading for every change of air temperature or voltage. With a given normal reading for a particular temperature and voltage, the reading will be 15% low with a shortage of refrigerant, 35% low with a restriction, 50% low with the valve jammed open in the compressor, and 15 to 20% high with air in the system or with a dirty condenser. If the starting winding does not cut in, the reading will be high; if it does not cut out, the reading will be very high; and, if badly shorted, it will be "sky" high. Table 6-1, which is for a particular 1/8-hp motor, shows the scale or relationship of the readings with various situations.

An early arrangement in refrigerator design was to install a separate door on the across-the-top freezer. This was an excellent practice because it prevented warm air from entering the freezer every time the food compartment door was opened. In addition, frosting was reduced (frost is only condensed air moisture), and less moisture-laden air was introduced. Unfortunately, this kind of freezer would not maintain zero or sub-zero cold; it was difficult, for example, to refreeze gelatin-base ice cream. Thus, the two-door, top-freezer style was introduced, in which the food compartment was free from frost, at least, and the freezer had less of it since its door was not opened very often. The food compartment had either a chill plate on the back or cold tubing on the top of its liner. In either case, when the compressor shut off, the accumulated frost would melt and the water would run down to a pan below, where the warmth would evaporate it. For this to hap-

TABLE 6-1 Characteristics of Some 1/8-HP Motors

Characteristic	Range with Capacitor			Range without Capacitor		
	100 V	115 V	125 V	100 V	115 V	125 V
Running wattage at						
70–90°F	105–128	108–138	114–144	105–128	108–138	114–144
90–110°F	125–155	133–163	139–169	125–155	133–163	139–169
Starting wattage						
Normal stall	900–1000	1250–1350	1525–1625	1700–1800	2200–2350	2500–2700
Stalled shorted capacitor	1700–1800	2200–2350	2500–2700			
Open started winding	500–550	700–750	850–900	500–550	700–750	850–900

pen, the switch had to make the bulb warm up to 37.5 degrees before the compressor would start again, resulting in a very long cycle. One of these cycle-defrost types can very well run for 20 or 30 minutes and shut off for the same length of time or longer.

The consequence was a greater "swing" of temperature in the food compartment than was common with the earlier types, some of which had a *very* short cycle. The builders had to make refrigerators that would work throughout the vast range of climatic conditions in the U.S. To make sure their units would stay shut off long enough for the frost to melt, they sometimes inserted a piece of plastic between the thermostat bulb and the chill plate, which was part of the evaporator. The plastic served as an insulator that lengthened the cycle.

The best way to defrost a freezer is to use a hair dryer to melt a hole through the frost. The surrounding metal will then warm up sufficiently for the frost to be knocked off with a piece of wood.

The modern thinking is that it takes so long to cool up a cycle defrost refrigerator after it has been defrosted that the frost-free refrigerator really does not use any more electricity.

There are two kinds of evaporators: soldered-on tubing and "bonded-on" sheet metal. In the latter style, one sheet is "embossed"; when it is bonded to another sheet, there are passages for the refrigerant to flow through. The soldered-tube type will sometimes develop a leak at the point of attachment because of the corrosive action of the flux used. Such a leak is extremely difficult to repair because of the restricted space, rough surface, and presence of oil. The bonded-on, or "rolled-on," type can also leak, and if the leak is caused by a puncture, it would seem worthwhile to repair it. If the leak is caused by corrosion, you can also repair it, but you probably will be fighting a losing battle. Leaks will keep on developing.

Be warned that epoxy takes a long time to set up. If the room air warms up in the meantime, it will create an internal pressure that will push gas right through the unsolidified epoxy, thereby leaving a hole at the very point you are trying to cover. To prevent this, the temperature has to be kept constant. A slight vacuum in the system is desirable, if possible.

The cycle defrost refrigerator may present a water disposal problem. Although the water is supposed to go to the disposal pan near the floor, it can run out under the door if the passage is obstructed. Sometimes there is a trough under the chill plate or coil, with a tube running down to the bottom, and other times the defrost water trickles down the back of the food compartment and out through a bottom drain. Obstructions are caused by algae (a form of plant life in the air) sometimes combined with food particles or gravy. Bottom drains can be cleaned by removing the "hydrator trays" (or whatever term the manufacturer uses) and taking out the part of the drain that is plastic and washing it. In no-frost refrigerators, more water must be disposed of. Clearing of drain water tubes will be discussed later.

7

The No-Frost Refrigerator–Freezer

Refrigerator-Freezer

The extent of frost trouble in a freezer depends on use, that is, it seems as if the so-called "deep freeze" (where the *entire* space is a freezer) is opened less than a small freezer combined with a refrigerator.

The refrigerant circuit pictured in Fig. 7-1 has several interesting features. The bottom layer of freezer coils has fewer passes because cold has a tendency to settle, requiring less cooling area at the bottom. The arrows show that the warm vapor is pushed into the superheat coil where a large part of the heat is removed. The warm vapor then goes into condenser coils in the refrigerator walls. The wall coils are in right and left sections; Freon is directed into each section. The liquefaction of the refrigerant is completed in the wall coils, and, at the "T" fitting, the Freon discharged from the two sections merges and enters the capillary line to complete the circuit.

Refrigerator-freezer combinations that have little or no frost have operational principles that are somewhat more complex. Complete diagnostic procedures for all the models that the various companies have marketed would require a huge book; we must content ourselves with some generalities.

Electric Defroster

In the no-frost refrigerator, defrosting is accomplished by the application of heat to the evaporator coil. This can be by way of an electric heater or perhaps a hot gas coil. The water runs down to a pan at the bottom, where it is evaporated, and this evaporation is combined with "sublimation" by the movement of air. This concept will be explained later.

Because electrical controls have moving parts, they are subject to mechanical difficulties. For example, a control might work perfectly for 99 times and then "stick" either open or shut; it might release itself and then return to normal operation for an unpredictable length of time. With sophisticated systems, the interrelation of the effect of a partial discharge of the sealed unit combined with erratic operation of the controls presents a

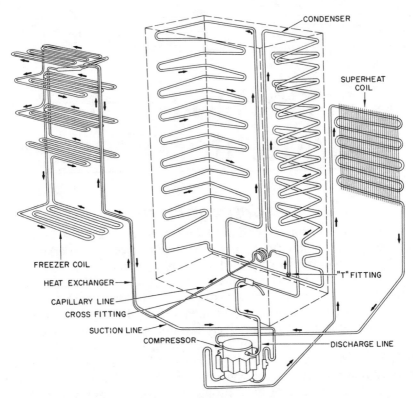

Fig. 7-1   Refrigerant circuit in an upright freezer (*Courtesy*, Frigidaire)

diagnostic problem that is not as simple as it is with a "basic" system. Because the purpose is to melt the frost quickly and hold the thawing of food to a minimum, the electric heater has to be of fair size with, say two-thirds the heating capacity of a flat iron. There has to be a timer to turn the heater on and the compressor off. There should also be a thermal-limit switch to shut off the power when a predetermined temperature is reached. Brand name makes incorporate this safety device, but some "cut-rate" refrigerators omit it. If the timer of one of the latter sticks in the "on" position, the freezer will become *very* hot. Most complaints, however, result from an excess of frost on the evaporator coils. To remedy this situation, you have to remove a liner panel and defrost with hot water. With the compressor running, leave the unit alone for a little while until the evaporator frosts up. Then, by studying the frost pattern, you can decide if the evaporator has discharged refrigerant.

The defrost timer will turn on the heater either on an elapsed-time basis or after a specified number of hours of compressor operation. In respect to the former, defrosting usually occurs every 8 or 12 hours. With an infra-red heater, a red glow can be seen through the air holes in the back. There have been those who have become alarmed upon noticing the glow after the refrigerator shuts off, but for a unit of this type, it is perfectly normal operation. After some number of minutes, the heater shuts off and the unit begins cooling again.

Hot-Gas Defroster

Figure 7-2 shows an evaporator tube that carries hot gas from the condenser at specified times. The hot-gas tube might not be as long as the cooling tube because the heating coils work more effectively, and not as much radiating area is required. In other words, the hot-gas tube might take short cuts and not follow the cooling coil exactly.

For defrosting, the timer energizes a solenoid that raises its plunger, opening an orifice and allowing the hot gas to flow into the outer tube. If the unit will not stop defrosting, it would be worthwhile to apply the test bulb wires to the solenoid terminals. If the bulb lights, indicating that the solenoid is energized, it appears that the timer is out of order—perhaps the clock stopped when the contacts to energize the solenoid were closed. If the bulb does not light, this indicates conventional troubles or that the hot gas was still going through the de-energized valve. This, in turn, means that

Fig. 7-2 Hot gas tubing for defrosting freezer

Fig. 7-3   Frost disposal during "off" part of cycle: (1) evaporator plate, (2) fans, (3) plastic tube, (4) trough, (5) pan, (6) condenser, (7) freezer air distribution duct, (8) freezer drain, (9) chill coils (*Courtesy*, Frigidaire)

sticky oil is binding the plunger or that the spring is broken. If the plunger is stuck, gentle tapping might free it.

Opening the passage into the defrosting coils reduces the high-side pressure and reduces the flow of Freon into the cooling coils. Because a little extra Freon is needed when both coils are in use, an "accumulator bulb" serves as a "storage place" to hold the extra Freon at times when the unit is not defrosting.

The other end of the hot-gas (defrost) coil connects to the suction line. When an increase in cabinet temperature is noted, its cause might be any of the conventional refrigerator malfunctions, or it might be hot gas leaking through the solenoid valve.

Another method of keeping the unit from frosting uses an evaporator plate that collects frost when the unit is running, but when the unit shuts off, the frost melts and drips into a pan located where the heat of some loops or "passes" of the condenser tube will evaporate it. In Fig. 7-3 we see water dripping from the plate into a plastic tube and running down to a pan. When the unit defrosts the same thing occurs in the bottom (freezer) compartment, which is, the frost melts and drips down to the pan as shown.

The temperature differential between the freezer and the refrigerator sections of a combination unit varies somewhat according to make, model, and operating conditions, but it might be 45° F. Thus, the temperatures could be 45° F in the refrigerator and zero in the freezer. If you check the temperature range of the switch, you will find it will indicate a higher temperature than the actual freezer temperature. That does not mean that the switch is not calibrated properly.

Actually the switch bulb, which is bolted to the sheet metal between the tubing outside of the freezer compartment, is affected by whatever heat comes through the insulation. Therefore, the switch is governed by the temperature of the metal to which it is bolted. The coils have to be colder than the "cut-off" point to make the sheet metal, with the switch bulb, the same temperature as that found inside the freezer.

To get a pictorial idea of what goes on in a unit of this sort, examine the time-vs.-temperature graph for a well known product (see Fig. 7-4). This situation would be fairly normal with moderate operating conditions, but what happens when the unit is operating under abnormal conditions? Suppose that the room is 95 °F. and a lot of cold food was removed and replaced with warm food (and beverages). In this case (see Fig. 7-5), after an initial drop of more than 20°, 2½ hours would be required to drop from 10 °F. down to zero. After that, the unit (unless more replacement of food took place) would assume a cycle of something like the one shown in Fig. 7-4.

Note that, in spite of the considerable variation of the freezer coil temperature, the temperature of the frozen food does not vary much, only a

Fig. 7-4 Time vs. temperature for normal operating conditions

Fig. 7-5 Time vs. temperature for unusual operating conditions

few degrees. Note also that, after a prolonged period of intense cold in the freezer as shown in Fig. 7-5, the cold-control switch shuts off the motor at a slightly higher temperature.

"Remote-Reading" Thermometer

In Fig. 7-6 we have a device used for sensing the temperature changes inside the cabinet while the door remains closed. This "remote-reading" thermometer has a cable (stored in the back when not in use) with a sensing probe. With the case resting on top of the cabinet (or wherever is convenient), this "sensor" can be placed inside either the frozen or fresh food compartment and the door closed on the cable.

Fig. 7-6 Single-range, one-probe, remote-reading thermometer

Figure 7-7 shows a more elaborate device of this type. It has two temperature ranges that permit readings from − 50° to 150° F. The cables, or probes, can be used at the same time. All you have to do is turn the selector knob to give a reading on the probe you wish to read. This practice of

Fig. 7-7 Dual-range, three-probe, remote-reading thermometer (*Courtesy*, Robinair)

using three probes is very handy when working on an air conditioner, because the probes can be put in three different places to check the circulation of air.

To return to trouble-shooting procedures, we may say that with a combination unit, a refrigerant shortage or a restriction will cause variations of temperature that can be plotted in a graph such as that shown in Fig. 7-5. The readings for such a graph can be obtained with a remote-reading thermometer and a clock.

"Sublimation"

These units sometimes make use of the process of sublimation in the freezer. Sublimation is defined as the process by which a solid changes to a gas without passing through the liquid state. This process occurs when a cake of "dry ice" evaporates. In our case, we apply the process by blowing air over the frost so that it evaporates without having to be melted first.

In Fig. 7-8 we have this taking place. Fan E pushes air through duct F and out opening 1. The air moves through the freezer into opening 1 and through duct 3 into the bottom space where it passes over coils 4. The cage-type fans (refer to Fig. 2-9) used for air circulation have self-contained, permanently lubricated motors, and it is believed that they can run for a great number of years without breaking down.

The no-frost refrigerator has its defrost heater in series with a thermal-limit switch. There is also a fan to draw air over the freeze coil, that is, over the tubes and between the fins, which are at an angle of 90 degrees to them. As we know, moisture in air will attach itself to anything that is cooler. In this case, the moisture becomes frost on the coil, but if for some reason the

Fig. 7-8   Air circulation pattern: Fan (5) pushes air through duct (6) out opening (1). It moves through the freezer and into openings (2) and through duct (3) into the bottom space, where it passes over finned coils (4) (*Courtesy*, Frigidaire)

frost gets so thick that the air cannot pass through it, the freezer and food compartment will begin to warm up. Furthermore, some radiational cooling in the freezer will be interfered with. In fact, most cases of high cabinet temperature are caused by an over-frosted coil.

To remedy this malfunction, first find out if the fan is running normally. You may have to push the light-switch button to do so. If the fan is stuck or moving slowly and making funny noises, it should probably be replaced. If it is OK, you must now determine whether the defrost heater is coming on properly. The defrost timer should have a shaft that can be turned with a screwdriver or with the thumb and forefinger. Manually advance the shaft until the compressor shuts off. Use a wattmeter to check out the heater; if it is working, the wattage will jump up, possibly to 600 watts with a resistance coil but somewhat less with an infra-red bulb type. If the heater does not come on, there must be an open circuit in the heater or the thermal-limit switch. If it does come on, either the timer is not advancing or the drain is obstructed.

The principle of operation of a timer is to have its motor shift an arm (see Fig. 7-9) to energize either terminal No. 1 or 2. (Your own timer terminals may not be numbered the same, of course.) Some timers have more than four wires, but only four are required for the defrosting function. If the timer motor is running, the heater will shut off and the compressor will start up shortly thereafter. Timers, unfortunately, have been known to "stick."

If everything else is found to be in perfect order, you may as well conclude that the timer is unreliable and replace it. In any event, the accumulation of frost and ice has to be removed, and to do that you have to gain access to the freeze coil. It will be found at the back of the freezer or beneath it. First you must remove whatever pieces of plastic or sheet metal may be inside the freezer. If there is an automatic ice maker, it too must come out. Then you must remove either the back or bottom panel of the freezer liner by taking out the sheet metal screws that hold them in place. The frost and ice may then be melted with hot water or whatever other forms of heat are

Fig. 7-9 Operation of the timer motor

TWO DIFFERENT METALS

PERMANENTLY BONDED TOGETHER

WHEN HEATED....

HIGHER EXPANSION OF ONE SNAPS DISC INTO REVERSE CURVATURE

(A)

CONTACTS

SPENCER DISC

OPEN

CLOSED

(B)

Fig. 7-10 Bimetallic thermostat

available. If the heater still functions, use it; a drop lamp or hair dryer is also a good bet. With the frost and ice removed, you may locate the trouble. If the heater did *not* come on, cut one wire and scrape a place bare on the other. (You will see either an oven burner type of heater or a long, thin glass tube.) For checking continuity, a battery-powered bulb tester used on cars will come in handy. Try to get one with a large bulb, which has less resistance in its filament than a small bulb and thus does not require as good a contact. A small bulb light, moreover, is hard to see except in a shadow.

In the process of defrosting, you will have caused the limit switch to open. By spraying liquid refrigerant on it, you can make it reset, but to do so in a freezer takes three or four hands at least. A good way to do it is to cut both wires and remove it. By holding the limit switch with the foot, it is possible to direct liquid onto it without assistance. By stripping some insulation from the wires, you can connect the car tester clamp to one wire and use visegrips to attach the probe to the other. If the bulb lights, it is OK. (Remember, always, that Freon can blind you.)

If there is an inoperative drain heater, it is likely that the wires have a bad connection where they are crimped on.

The limit switch has a bimetal, snap-action disk (see Fig. 7-10). Heating causes the bimetal to warp and open the contacts. If a solid state device is used, the function is the same. In a top-freezer model, the drain water either goes to the bottom through a pipe or it runs down the back of the food compartment liner to a hole in the bottom, as described in the previous chapter. In addition to algae, the pipe can be obstructed by pieces of foam insulation that have broken loose and been carried away by the flowing water. One way to clear the pipe is to put a ¼-inch flare connector on the end of the refrigerant hose. Attaching the hose to the end of the (plastic or rubber) pipe will not give a perfect seal but allows for a sufficient "blast" of refrigerant into the pipe to dislodge almost anything except ice. You have to hold your hand around the hose, of course.

As far as *ice* is concerned, you can puncture it with a hot soldering iron. Defrost water has to go somewhere; if it does not go down the pipe, it will go into the food compartment and run out under the door, creating puddles that can make a housewife very unhappy. Note that replacing a

Fig. 7-11   Typical air flow and water disposal in a top freezer

drain pan that has thoroughly rusted through may require a great deal of dismantling.

Sometimes there is a simple solution to the defrost water problem. If the food compartment liner has a "sump" in the bottom and *foam* insulation, you can take a piece of 3/8-inch O.D. copper tube, flare one end, and drill a 3/8-inch hole in the sump, making sure that the drill doesn't penetrate anything else so that the water will drain into the pan. Then simply put gasket sealer around the hole and push the tube in, and the water will go into the pan as it is supposed to. If you drill into *fiberglass* rather than foam, it will wrap around the drill, and you will not be able to pull it back through the hole. Two holes are required, of course—one in the sump and another in the shield over the motor compartment.

The side-by-side type of freezer operates on the same basic principles,

Fig. 7-12   Typical air flow and water disposal in a side-by-side freezer

except that the air flows partly sideways. A variation is the three-door type, which allows ice cubes to be removed without disturbing the stored food. Figure 7-11 shows air flow and water disposal in a top freezer and Fig. 7-12 in a side-by-side freezer. The no-frost type usually runs on a small temperature differential between states of "on" and "off."

In considering the difference in cost of running frost and no-frost units, it must be remembered that frost acts as insulation and consequently

CONDENSER | FAN HOUSING MTG. SCREW | CONDENSER FAN MOTOR | MOTOR PROTECTOR

CAPILLARY LINE | CLAMP & BOLT | LINE GROMMET | SUCTION LINE | INSULATION

Fig. 7-13 Condenser fan motor mounting

that the compressor has to run more frequently. If people would defrost nondefrosting units more often, their relative efficiency would still hold up pretty well. As it works out, however, many people allow a great accumulation of frost and ice to build up before they remove it. In frost-free units, on the other hand, there is not enough frost at any one time to matter much.

A frost-free unit either has the condenser on the back or a fan in the bottom to pull the air in on one side and push it out on the other (see Fig. 7-13). Refrigerators are now often boxed in with cabinets; for those with a back-mounted condenser, that is a very bad idea, since carpenters hardly ever allow enough space for proper ventilation. Some people have even had to make a hole in the back partition. If the inside cools well enough, the owner may take the attitude, "It works, doesn't it?" The trouble with poor air circulation is that the compressor will run more often than it needs to, and the excessive heat in the motor compartment (at the bottom) will carbonize the oil. As a result, the life of the unit will be shortened by years.

The no-frost refrigerator-freezer cold-control switches are likely to have the bulb in an air passage. There is a wide variation in them in terms of operating ranges and differential.

8

Other Tests, Procedures, and Comments

The heavier unit has a 208/230-volt or three-wire input. This equipment operates in the same manner as the 115-volt equipment, but the condenser might be water cooled and the restrictor is likely to be replaced by an expansion valve.

Expansion Valve

An expansion valve is needed to handle the larger volume of Freon that passes into the evaporator in these units. There are several types of expansion valves, but, on household units, it would probably be of the thermal type, which means, simply, that it has its own gas-filled bulb, tube, and bellows that senses a temperature change and causes a mechanical action, which in this case, is the opening or closing of the intake port in the valve. In Fig. 8-1, we can see that expansion of gas in the bulb will result in pressure being exerted on the diaphragm, which in turn transmits that pressure to the plunger cap. This compresses the spring, and pushes down the plunger shank, causing the plunger body to uncover the intake port.

On relatively heavy household or light commercial equipment, a solenoid could move the plunger to accomplish the same result. But, with either type, a drop of water falling on the junction of the needle point and the orifice, can freeze, and, thereby, shut off the flow of Freon through the valve.

When a unit with an expansion valve will not cool, it is worthwhile to pour hot water on the valve. This will melt any ice that is present. If that does not help, try tapping the valve. If it has a solenoid, use the test bulb to see if it is energized. If the test fails, the solenoid coil may have a broken wire. You can test a solenoid by taking it off and putting a nail in it and energizing it. If there is a magnetic field, the nail will jump in a most interesting fashion. Of course, lack of cooling might stem from lack of Freon. Whether or not this is the case can be determined with a pressure reading, which we will discuss later.

DIRECTION OF
PLUNGER MOVEMENT

TO BULB ON SUCTION LINE

DIAPHRAGM

PLUNGER CAP

PLUNGER SHANK AND
SPRING INTAKE PORT

INTAKE PORT

FREON FLOW

PLUNGER BODY

FREON INTAKE

Fig. 8-1 Thermal expansion valve

Remote Condensers

As for condensers, it is apparent that, on an air conditioner at least, you can not very well blow the hot air out the window if the unit is in the middle of the house. In this case, the condenser has to give its heat to water, which, in turn, gives its heat to the air. Or, at least, that is the usual arrangement. Now, you could just let the warm water run down the drain, but, in most places, water costs enough so that it can not be wasted. Thus, the cooling water is pumped around in a "closed circuit" with the heat being given off to the air in a type of radiator. This "radiator" is likely to be on the roof, but wherever it is, the water will not flow without being pumped. So, if there is something wrong with the pump, you are obviously going to have a warm house.

A common difficulty with these water-cooled condensers is that a "hard water" condition results in the accumulation of scale on the tubes. The scale is likely to form over a period of years, and the unit slowly loses its efficiency. When scale formation is the problem, quite a bit of the scale can be removed by opening the tank and scraping it with a wire brush. Of course, there are chemical ways of removing scale. The Calgon Company manufactures some preparations that are intended for this service. One is claimed to "anticipate" scale formation and prevent it; the other is to be used after scale formation has already taken place.

In the old days of "open" units, the standard testing procedure was to take pressure readings on fittings provided for this purpose. On a modern sealed unit, however, provisions are not made for pressure readings. The volt-wattmeter is used, and it, combined with the use of the sense of touch (relative to the temperature of the various parts of the system) and hearing, will provide a diagnosis in all but the most complicated cases. Electrical testing should be done first because taking pressure readings involves cutting into the system (cutting through a tube or punching a hole in it), and that should be done only as a last resort. Of course, the equipment might have a fitting for a low-side reading, which, of necessity, would require the use of a Robinair charging tool.

Compound Gauge and Manifold

If pressure readings are necessary, a manifold assembly has been devised that has two gauges mounted on it. One gauge, for the high side of the system, reads from zero to, say, a few hundred pounds per square inch (psi). The other gauge, for the low side, gives a "back-pressure" reading somewhere between say, 150 psi and 30 inches of vacuum.

The manifold also has three intake ports (¼ flare) with a valve stem on each end. With the valve at one end open and that at the other end closed, a high-side reading can be taken, and, when the valves are reversed (the open one closed and the closed one open), a low-side reading can be taken. The third intake has a purpose, but, when the manifold is used simply for testing, it can be capped.

One situation in which the manifold will indicate that something has gone wrong inside the compressor is when the pressure on both sides of it is the same. For instance, if the high side, or "head" pressure is 75 psi and the "back" pressure is 75 psi, obviously something is wrong with the compressor.

Temperature Vs. Pressure

In other situations, the readings can give "presumptive evidence" of a malfunction. To gain an insight into how it works, let us consider how the figures on a refrigerant temperature-vs.-pressure graph are obtained. Let us say that you have a flask of Freon-11 in a room in which the temperature is exactly the same as the vaporization temperature of the Freon (in this case, 75°F.) If the flask is full and you connect a compound gauge to it, you can vary the temperature and get the same readings as on the graph. That is, if the room was heated, the Freon would expand and increase the pressure in the flask. How much the pressure increased would depend on how much the temperature increased. At 90°F. it would increase 5 psi and at 100°F. it would increase 8.9 psi. At 0°F. it would decrease to 24.7 inches of vacuum and at −20°F. it would decrease to 27 inches of vacuum.

Of course, it is unlikely that you would find Freon-11 in household equipment, but we are using it as an example to show that with a room temperature of 65 °F., the normal readings would not be the same as with the room at 95 °F. In addition, a shortage of refrigerant would make a difference. If there is no other way to reach a conclusion, the unit can be discharged of the refrigerant, and then it can be recharged with the amount specified on the unit information plate. You must take care that air does not enter into the system.

We might see the usual high- and low-pressure relationships by relating some sample readings to a particular refrigerant and room temperature. With Freon-12 in a room at 90 °F., the head (high-side) pressure should be 100 psi or higher, and the back (low-side) pressure should be 8-12 psi. If the head pressure is 100 psi and the suction side has 25 inches of vacuum, the unit is restricted. With a high suction reading and rather low head pressure, it would be worthwhile to add some Freon and see if normal readings will result.

Pressure Readings

Taking a head-pressure reading is simple—all you have to do is screw the charging tool onto the charging stud (using the proper copper gasket) and attach a pressure gauge directly to the tool with a flexible tube with knurled "finger-tight" tightening attachments on the end. At least, that is what you can do if the compressor is fitted with a high-side charging stud.

It is not unusual to encounter a compressor that was charged through a "process tube," which is just a length of tubing extending out of the compressor housing. The person doing the charging just crimped or flattened the tube, cut it off, and soldered it when the charging was completed. To get a reading from a compressor with a process tube, you have to use a fitting the same as you would use to take a back pressure reading. Now we will see how this can be accomplished.

If you decide to use the following ideas, take warning: by punching holes in, and otherwise tampering with, a sealed unit, you run the risk that moisture will get in and carbonize the oil. This would not necessarily happen right away; a leak will eventually cause the suction side to develop a vacuum, which, in the case of a low-side leak, will cause air (moisture laden, more or less) to be drawn in. The eventual outcome will be a stuck compressor or a serious restriction. If the unit is in warranty, puncturing or cutting the tubing will void the warranty.

At one time it was customary to install a "T" block (see Fig. 8-2) as a means of recharging a sealed unit when a charging stud was not provided. This is a type of depressor fitting; to open a passage, it is necessary to use a special hose that will depress the valve stem. When the hose is removed, the pressure of the gas, assisted by a spring, pushes a valve plate against a neoprene washer to close the opening.

Fig. 8-2 "T" block fitting

A suction-line or high-side reading also can be taken with a device that clamps to the tubing and has a neoprene block that bears against the tubing (see Fig. 8-3). The neoprene has a hole in the center through which a pointed shaft passes. Turning a handle on the end of the pointed shaft moves the point away from or toward the tubing. After being clamped on, the device is put into use by turning the handle to cause the point to puncture the tubing. (The gauge should be connected, of course.) With a passage into the tubing thus opened, a reading can be taken by backing the point out of the passage. After the reading, the passage is closed by turning the shaft so that the point

Fig. 8-3 "Clamp-on" fitting

Fig. 8-4 Charging kit with manifold, gauges, connecting hoses, and Freon holder

fills the hole. Once you have installed this fitting (which is available with a depressor valve) be sure to cap the connection.

Replacing a Component

A component replacement requires a vacuum pump. Specially made ones are available, but any motor-compressor will serve. One can be obtained from a unit that has been scrapped for some other reason. The suction line would have to have a ¼-inch flare fitting soldered to it.

A supply of Freon is also required. A professional would use a charging stand that has a "sight glass" that is marked in ounces of Freon (see Fig. 8-4) for withdrawing the exact amount of refrigerant. However, the job can be done without a charging stand. Fifteen-ounce cans can be coupled together with tubing to supply the required amount, whatever it might be. Of course, you have to be careful not to overcharge the unit with this arrangement.

To set up the equipment (see Fig. 8-5), connect the charging attachment to the middle port on the manifold. Connect the vacuum pump to the compound-gauge end of the manifold and the Freon supply to the remaining port.

When you are ready to cut the tubing, cut, couple, and evacuate as

Fig. 8-5   Equipment connections of charging kit

quickly as possible so that the interior of the sealed unit will not be exposed to the air any longer than is necessary. With the tube ends coupled together, start the pump. The charging stud on the compressor should be left closed. The idea is to see if the fittings are leaking. Let the pump run for five minutes. The vacuum reading should be about 27.5 inches at sea level. Stop the pump and let everything stand for five minutes. The vacuum reading should be the same at the end of the period. If not, you will have to locate the leak and repair it. One source of leakage is the manifold valves. The stems have to be turned either all the way in or all the way out.

Now, open the charging stud by turning the fitting with the handle on the charging tool, and repeat the same procedure as outlined above. In this case, the evacuation and testing times are longer: Pump out for 20 minutes and hold a vacuum for 20 minutes. A perfect vacuum is difficult to attain, but the reading certainly should not drop below 27 inches. If the vacuum holds, close the valve to the pump, and let in the Freon. After the correct amount of refrigerant is in, close the charging stud and start the unit motor. Let it run for a few hours to determine that it is not over or undercharged.

Figure 8-5 shows how the various connections are made. Of course, it does not *have* to be done this way, but, if you use this procedure, you should have, at each stage of the program, the state of affairs shown in the table at the top of the facing page.

At the end of charging, close all valves and disconnect. Note: Valve for line to charging stand has to be open to evacuate air from the line. If the charging stand valve is leaking, obviously the valve where the line connects to the manifold will have to be shut.

	Valve on charging stand	Vacuum pump	Valve on vacuum pump	Charging tool	Valve for line to charging stand
First evacuation	Shut	On	Open	Shut	Open
Test fittings	Shut	Off	Shut	Shut	Open
Second evacuation	Shut	On	Open	Open	Open
Test for leaks	Shut	Off	Shut	Open	Open
Charge	Open	Off	Shut	Open	Open

When replacing a component part of a sealed unit, you should install a "dryer" on the high-side line. The technical term for it is "conical screened molecular sieve dessicant container." It is just a bulb with a tubing connection at each end, but it has to be installed so that the Freon will move through it in a specified direction. The dryer is filled with granules of a chemical compound. The Freon has to pass around these granules that have the property of attracting and seizing whatever water is mixed with the Freon until they become saturated. At the outlet is a cone-shaped screen. Any solder, dirty oil, etc., that comes along is supposed to be pushed down to the base of the cone and not obstruct the subsequent passage of Freon. Note: If a depressor valve is used, the hose connections can be changed to provide for a shutoff.

Refrigerant Properties

At this point it would be worthwhile to briefly mention certain properties of refrigerants. If Freon is subjected to intense heat (or on fire) it will break down chemically and become a poisonous gas. The two other refrigerants that were popular years ago and still may be encountered are sulphur dioxide and methyl chloride. The former has the properties of tear gas. If any amount of it is in the air, you will need a gas mask. To find points of leakage soak a rag in ammonia and hold it at suspected leakage points. Direct contact of the two will result in the formation of a heavy cloud of white smoke. As for methyl chloride, people using a flame type leak tester have found that it will burn. Furthermore, it can undergo a chemical transformation through the years so that, when evacuated, it can combine with oil to make a rather dangerous explosive.

Testing 230-Volt Equipment

You can test a 208/230-volt circuit in the same way as you would an ordinary 115-volt circuit. You can insert the probes in a three-prong receptacle and find that, between one of the live probes and the ground, the bulb will light dimly; between the two live probes, on the other hand, it will burn brightly.

(A)

(B)

Fig. 8-6 (A) A 115-volt circuit, and (B) component assembly of 115-volt circuit (*Courtesy*, Frigidaire)

A word of caution: *Before attempting to test or repair higher-voltage equipment, you would do well to secure the advice of someone who has had experience in such matters unless you are completely sure of what you are doing. Working on equipment of this sort presents the possibility of serious, even fatal, injury.*

You might be using a schematic which might be a bit confusing. Figure 8-6A is such a schematic and Fig. 8-6B shows the actual appearance of the parts and connections indicated by the schematic.

There are many ways of connecting motors with and without capacitors and relays, but in one particular situation, when you trace the current through the switches and protectors, you will find the motor is connected as shown in Fig. 8-7. Because there are three wires that are connected to the motor, it follows that you can use a three-wire test cord and bypass everything that you are not directly concerned with.

Fig. 8-7   One method of connecting a 208/230-volt motor

Phases

There is considerable confusion when reference is made to motor phasing. Just what does "phase" mean? The phases originate in the power company's generator. For explanatory purposes, we may assume that the generator stator (outside or stationary part) is divided into three segments, each with a wire leading from it. The rotor makes a revolution in one time unit. Suppose that there is an "X" on one of the two rotor segments, which is the one that rotates to produce the power you are using. Let us label the stator segments A, B, and C (see Fig. 8-8).

Fig. 8-8   A typical generator

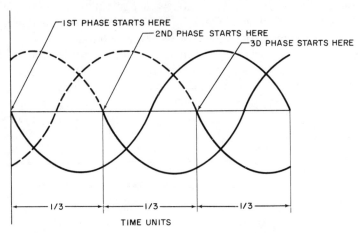

Fig. 8-9   Sine-wave representation of three-phase current flow

Each time the rotor segment marked "X" passes one of these stator segments, it sends out a cycle of ac current. (The electron flow goes out, reverses itself, and then reverses again.) Now, because the rotor segment has to go past A to get to B, it follows that the B cycle is going to start a third of a unit of time later than the A cycle, and the C cycle is going to start two thirds of a unit of time later than the A cycle. "Staggering" the cycles in this fashion is called "phasing." The time unit referred to would be 1/60th of a second for 60-Hertz current. Thus, the wire from A would begin a cycle, and 1/3 of 1/60th of a second later the wire from B would begin a cycle, and so on. In Fig. 8-9 we see a representation of the current flow. Each phase is represented by a line that curves and returns to its axis according to the reversal of electron flow.

208/230-Volt Systems

Whether 208 or 230 volts is delivered to your home is governed by the way the leads on the distribution transformer are connected. There are several different ways of connecting them. With one method, you get 115 volts on an energized leg, and, using two of them, you get 230 volts. With another method, using two 120-volt legs, you get something less than twice 120. In this case it is 208 volts.

Figure 8-10 shows the possible connections on a four-wire, three-phase service. Now, one of these energized wires or legs can not do any useful work unless there is a circuit back to the generator (through the transformers). So, when one leg is used, the current has to flow through the so-called ground wire, and you get 115 volts.

In a 208/230-volt system, the two legs form the circuit, and the ground

Fig. 8-10 208-volt, four-wire, three-phase service

wire serves mostly as a way for voltage surges (such as lightning) to be carried away without damaging the equipment inside a building. On an air conditioner, the ground wire is attached to the chassis, and, somewhere on the other side of the receptacle, it would be connected to, possibly, a water pipe.

An ordinary three-wire system is called "single phase" because you can make a circuit with either phase and the ground wire. There also is a three-wire system in which all three of the wires are "live" and are connected to the motor. These "three-phase" motors are used on heavy equipment, with which we are not concerned. Having two different windings in a motor stator is called a "split-phase" arrangement.

Ohmmeter Test for Continuity

Perhaps the easiest way to check the continuity of a motor is with an ohmmeter like that shown in Fig. 8-11 or a volt-ohmmeter like that shown in Fig. 8-12. There should *not* be a circuit in a split-phase motor between the

Fig. 8-11 An ohmmeter (*Courtesy*, Robinair)

Fig. 8-12   A volt-ohmmeter (*Courtesy*, Robinair)

casing and a terminal. If there is, it means that an internal wire is rubbing on the casing and the insulation has worn off the wire, that is, it is shorted.

As for the windings, each one has a specified resistance in ohms. A handbook devoted entirely to motors might have a chart that would state, or give a close indication, of the resistances of ohms (Ω) of the particular motor that you were testing.

An ohmmeter test is made with the terminal wires removed. The meter has a little battery in it that sends out a steady stream of electrons (a direct current). If you run the current through a coil of wire, the dial will give a resistance reading in ohms. Direct-current resistances in series add up directly. (Alternating current is another matter.)

Figure 8-13 is a schematic of the windings of a motor with some resistance values assigned to them and the ground or common wire. (You would not actually have the values given; they have been chosen for explanation purposes.) Now, with the probes on black and red, the reading should be 2 plus 3, or 5. By the same token, black and white are 3, and red and white are 4. If the windings were shorted so that they were, in effect, connected together, the reading would be something else, and, if the wire were broken, there would be no reading.

Fig. 8-13   Typical motor windings

We have said that a motor-compressor can be started with a test cord. That is true, but some motors in sealed units have four terminals. In this case, the fourth one leads to a relay solenoid, which is only a different way of starting the motor. The fourth wire is on a third, or auxilliary winding. When the motor approaches the rated speed, the generator effect spoken of previously (the so-called counter electromotive force) will energize the solenoid so that its plunger will rise. The plunger is attached to a disconnecting device, and that opens the contacts.

Other Comments

To return to the matter of connecting the tubes when doing a component replacement, the present procedure is to solder them. At least, that is the usual practice with small units. Soldering is done with silver solder and often spoken of as "brazing." To connect two tubes of the same size, you need a "swedging" tool. (We are speaking of the suction line at this point.) This tool is a punch with a small end of perhaps ¼-inch diameter and with one sloping shoulder going to the 5/16 size, another to the 3/8 size, and so on. The idea is to drive the tool into one tube, thereby enlarging it enough for the other to be slid in—a pretty hard trick to pull off single-handed. The best way is to use the clamp block for a flaring set, put it on the tube being enlarged, and have someone else hold it.

As for the capillary tube, the way to cut a small one is to score it first with a three-sided file and then break it off. Don't try to clean the burr off the end. If a speck of metal happens to get in the tube, it will be carried to the restrictor. In air conditioners, some tubes can have an inside diameter as large as 1/10-inch. To cut these, you could use a tubing cutter.

At any rate, you will need a slightly large piece of tubing to slide over the two ends so that you can solder the ends of the larger piece. There is going to be a lot of pressure, and the solder job has to be perfect. A single drop of solder drawn into the end of one of the tubes being connected can run over the cut end and plug the opening.

Now let us consider the logic involved in the diagnosis of a refrigerator problem. Suppose that friends have bought a new no-frost model and report that drops of water are falling from the top of the liner of the food compartment. When you go to take a look, you find that the kitchen is hot and steamy. Drops of water continue to form on the liner top even after you open the door. Finding the fan OK and the bottom of the drip tray moist, you conclude that the defroster is working. After taking everything out of the freezer and removing the panel, you find hardly any frost on the evaporator coil and assume that it has just defrosted. The condition of the unit would appear to be normal. With cooling equipment, however, it is never wise to *do* anything until you have run through your chain of thought again to make sure there is no flaw in the logic.

In this case, there is. Considering the humidity and the moisture in the refrigerator, there should have been some water in the drip tray, even a fraction of an inch. If the tray is merely moist, it is apparent that the coil is not collecting much frost. Even if the frost is evenly distributed, a moment's reflection makes you realize that if the freezer were colder, there would be more frost. You must now decide *why* the freezer does not get colder. By invoking the principle of heat transfer, you recall that the refrigerant has to give off heat in the condenser after it has carried it off from the freezer. This realization brings you to the crux of the matter: Your friends' old refrigerator had a fan in the bottom and this one has the condenser on the back. Now you notice that there are cabinets on both sides of the unit and only about three inches of space above it to the overhead cabinets. To make matters worse, there are ornamental boards on both sides. You therefore decide to pull the refrigerator out of its cubby to see what will happen. To do so, you put your feet on both sides of it and pull on the crosspiece at the bottom. As soon as you pull it out about 20 inches and stand up, a blast of air hits you in the face as if you had opened an oven door. Then when you check the interior of the unit, you see that it is reverting to normal operation.

Obviously, what was called for was a better circulation of air. Circumstances alter cases. In this case, there was a high ceiling and plenty of space over the cabinet. One man's solution was to remove the plywood shelves and top of the cabinet, leave its frame and doors intact and raise the refrigerator on blocks at a 5-inch distance from the wall. Not much else could be done in a rented apartment where appearances had to be maintained.

Expansion valves can be a problem. Some of them are set for 6° to 10°F. and cannot be adjusted for the superheat that is created when the refrigerant warms up going through the evaporator. Some have an adjustment screw that can be foolishly twisted too far, allowing liquid refrigerant to enter the compressor. Compressors are not built like water pumps, and their outlet ports are small. If liquid gets into the cylinder, it will, at the very least, smash the piston.

Design problems for heat pumps include liquid flooding and burnouts. To prevent flooding, extra heat on the outdoor coil has been suggested. As for burnouts, it must be remembered that motors have to have oil and refrigerant flowing through them to carry away the heat. With reduced flow, they can get so hot that the insulation will char. If the oil is nearly black, smells as if it were burned, and tests positively (with litmus paper) for acid there has been a burnout. Cold startups can aggravate the flow problem, and, in this case, electric compressor heaters have been found to be helpful.

Let's consider air circulation again. Every room has its own natural circulation pattern. The fans of some air conditioners put out a blast of air that is all out of proportion to what is necessary for this natural circulation. The natural movement of air according to temperature—hot air rises, cold air settles—is called *air gravity*. Coal furnaces used to keep whole houses comfortable with gravity circulation; no fan was needed. Now let us relate this phenomenon to a chill coil in a furnace duct. When the heat is on, the hot air goes up the pipe because that is its natural tendency, and the fan only gives it an assist. When there is a switch to cooling and the air moves in the duct in the same direction as it did for heating, it is working *against* gravity. A decrepit furnace blower will therefore be more obvious when used for cooling than for heating.

Causes of furnace blower malfunction include a slipping belt, a pulley slipping on the shaft, a binding bearing, or a motor bearing so worn that the rotor rubs on the windings. Aside from that, there is the *dirt* problem. Dirt that builds up on fan blades and coils can often be removed with detergent and a spray of water (be sure to wash off every drop of detergent). Steam cleaning can be used as a last resort (some filling stations keep a steam cleaner on hand for engines).

A house that is hard to heat will also be hard to cool, and for the same reasons. Heat travels by *conduction*—which is the direct transfer of heat through a substance or between objects in contact with one another—or by *convection*—which involves the movement of air—or by *radiation*. Infrared pictures have shown that a cement wall will lose more heat than anything except an open skylight. The heat goes through it by *conduction* and is given off by *radiation*. When you try to cool the structure, the same process takes place in reverse. Other materials besides cement and cinder blocks conduct heat—plaster, for example.

There was a time when two-story frame houses were built by running the 2×4's from the sills to the eaves and spiking the upstairs floor joists to the studs. As a result, the spaces between them acted as "chimneys." This type of construction not only increased heating costs, it was also a fire hazard since flames could easily shoot right up to the roof. (Eventually, carpenters began putting "fire break" boards between each pair of studs, although this practice interferes with insulating requirements.)

Making a house air-tight can get a furnace in trouble since it needs air for combustion. A restricted air supply creates a dirty flame. At best, the furnace will be inefficient, and at worst, it will be a poison gas generator.

Since R-22 gets hotter in the condenser and thus loses heat to the air faster, it may be asked why it is not used in everything. The trouble is that the pressures involved are higher, and it therefore costs more to make the equipment. High-efficiency units are expensive because they require a multitude of fins on the coils to improve the heat transfer.

If a sticky valve is the cause of a malfunction, it is possible that a little Marvel Mystery oil will loosen it. Since this oil has the property of loosening any thick oil that has become lodged on a valve seat, it is worth a try. It is inserted in the hose on the charging fitting. Loosen the nut at the bottom, and put the oil in the hose with a medicine dropper. Then push the oil all the way in with refrigerant.

For decades repairmen have been using a form of alcohol (synthetic anhydrous methyl) to prevent moisture from freezing in the capillary tube. Gibson used to tag condensers with a warning that alcohol had been installed at the factory and should be replaced if the unit was discharged. The specification for a refrigerator was the amount of alcohol contained in a 2-inch length of ¼-inch tubing. The latest word from DuPont is that alcohol will contribute to the eventual breakdown of the oil.

Let us take up the subject of compressor failure by considering the case where a motor has become so hot that the insulation has burned off the windings. It is obvious that there is heat dissipation from the motor-compressor shell, but the major part of the heat is carried away by the circulating oil and refrigerant, which then goes to the condenser for removal of the heat by air blowing over it. Of course, it seems contradictory to make a condenser even hotter, but that is the way the machine works. The truth is that the condenser fan is hard pressed to provide enough cooling in extremely hot weather. An outdoor unit may have direct sunlight shining on it, or an indoor unit may have restricted air circulation. When a burnout occurs, it may be speculated that the insulation had been a bit charred on numerous occasions and that a fan failure or low voltage finally broke the camel's back. It is reasonable to ask why a proper protector was not provided. If a protector is buried in the windings, the trouble is, of course, that there is no way to get at it if it fails. A compromise is the "knife and sheath" arrangement where a metal "sheath" extends into the windings and the "knife" is a sensor that can be withdrawn from the outside. At any rate, char will cause a reaction in the oil that makes it progressively darker. The oil is light brown when new, and heat and moisture discolor it. You can get oil out of a system by letting the refrigerant blow back out of a charging fitting. If you want to study the oil drops, catch them on a piece of glass or a saucer.

Another problem is "floodback," which occurs when liquid refrigerant enters the compressor. If the evaporator fan stops or air flow is blocked by a dirty filter, it is possible for the liquid to pass through without vaporizing. What happens first is that the refrigerant pushes the crankcase oil away from the oil pump inlet. Any liquid that continues to flow in will go into the cylinder intake ports. If any of it gets into a cylinder, something will break. To combat this possibility, the *suction line accumulator* was devised. It is a sort of tank in which the liquid has to make some turns, thereby getting a chance to vaporize (the accumulator thus works somewhat like the muffler on your car).

Yet another condition is called "oil migration." It was long known that heat pump compressors were being damaged by cold startups, but nobody was sure why. Then Tecumseh made some parts out of tempered glass to see what was happening. They put a window in a motor-compressor housing. It turned out that the cold oil was absorbing a tremendous amount of refrigerant, forming an emulsion. It is necessary for the oil to have the ability to seize drops of refrigerant and carry them through the system as a means of cooling the motor, but the "bonding power" is so strong that, on a cold shutdown, the oil in the crankcase will combine with *all* the refrigerant pushed out of the accumulator by the pressure in the evaporator. When the pool of oil in the "pan" emulsifies, the mixture can occupy several times the volume of the oil itself. When the compressor started, there was violent frothing and the oil-laden emulsion was pretty well pumped out into the condenser pipe. The chilled oil did not make the circuit back to the compressor as readily as warm oil, and it was theorized that the bearings were being damaged by lack of lubricant. The remedy was the "compressor heater," a sort of heat tape that goes right around the compressor shell. Some have been added to units already in the field and were wired so that the heater could be shut off when not needed. If you have one connected in this way, you ought to turn the heater on several hours *before* the heat pump will be used and make sure that it is working.

Some refrigerator units have an air space where the foam insulation is supposed to be, that is, on the sides at the top rear. If it has a top freezer, the refrigerator will sweat on the outside, and after a while it will rust. You can prevent this from happening, with a kit intended for putting foam insulation around windows, but you will have to put on a longer tube. A rigid tube of sufficient length can be pushed right through the existing foam to reach the void.

If you install a clamp-on charging fitting, try to have the hose connection point up. If you put it on the bottom, some oil may run out and make a mess.

It is hard to imagine a refrigerator catching on fire, but it is possible. When the defrost timer is located at the bottom front center and makes use of a two-piece plastic connector block, water can get into the connector and cause arcing and burning of the plastic. If there is blanket insulation above this area, it may be covered with lint (cotton and synthetic fibers) and saturated with grease from food vapors—all of which will burn. Water can come from the pan if it overflows or from under the door.

In refrigerators imported from Europe, the cold control switch has a different shape than the switches made here and does not mount in the same way. If the switch has to be replaced, you may have a considerable problem. You may have to find a plastic box to hold it and then mount this box inside the cabinet liner.

The motors in single-phase hermetic compressors may be classified as RSIR, CSIR, CSR or PSC—terms that we shall now explain. When a winding is energized, it creates a magnetic field, but since the motor has no mind of its own, it has to be "told" which way to start turning. For this purpose, another winding is required in which the magnetic field does not occur at exactly the same instant as in the first. Motor design takes account of the fact that induction lags and capacitance leads. Induction is a function of the amount of wire and the insulation on it. (The starting winding is made differently.) The effect of a capacitor is to send out a flow of electrons that does not coincide with the phase coming to the motor. Causing a lag or lead is "splitting the phase." Visualize three runners named for the electrical functions. Capacitance comes in first, Resistance (pure phase) comes in second, and Induction brings up the rear. If they were dragging something, three of them could do it better than two.

The Resistance Start/Induction Run (RSIR) motor is less costly to build than the others and is commonly used in sizes through ⅓ hp. A relay removes the starting winding when this motor reaches normal speed. The Capacitor Start/Induction Run (CSIR) motor is like the RSIR but with a capacitor added; it is used in sizes through ¾ hp. The Capacitor Start and run (CSR) motor has good torque and is used for applications through 5 hp. The relay cuts out the start winding and leaves the two capacitors in use for the running winding. Study a wiring diagram to see how the series and parallel circuits are used. The torque is not as good on the Permanent Split Capacitor type. It is used in sizes through 5 hp.

9

Freezing With Fire— The Gas Refrigerator

Absorption

A book on cooling equipment would not be complete without devoting some space to so-called absorption equipment. Absorption refers to a chemical process that uses heat to produce cold! Absorption is the basis of the "gas" refrigerator. All of the domestic gases have been used for fuel as well as the electro-resistance coil, and kerosene-fired units have been used in remote places where electricity was not available.

The chemistry was worked out in a laboratory in the 18th century, and it was applied quickly. Coal-fired units were made that worked after a fashion. Then, in 1857, Pierre Carre patented an improved design, and others proceeded to secure patents by the dozen, on both sides of the ocean.

The chemistry has not changed much through the years; the constant redesigning was to secure better utilization of the fuel and more dependable operation. How does it work? Here is what the Morphy-Richards Company has to say about their Astral refrigerators (see Fig. 9-1):

"The sealed unit contains water-ammonia and hydrogen under pressure, and cannot be repaired in the field. The heat generated in the boiler causes ammonia vapor and water vapor to be expelled from the solution and rise into the rectifier. The water vapor is condensed in the rectifier and runs back into the boiler. The ammonia vapor passes through into the condenser tubes where it is liquefied by air-cooling. The liquefied ammonia is gravity fed through the ammonia duct to the top of the evaporator.

"Precooled hydrogen is introduced into the evaporator by way of the gas heat exchanger. In the evaporator the partial pressure of the ammonia is considerably reduced due to the partial pressure of the hydrogen (Dalton's Law of partial pressure). This results in the evaporation of the ammonia and consequent absorption of heat through the walls of the evaporator. A suitable cover is fitted over the evaporator to allow the maximum possible absorption to take place.

"The heavy mixture of hydrogen and ammonia vapor passes down through the gas heat exchanger into the lower end of the absorber where it

Fig. 9-1 Absorption type cooling unit (*Courtesy*, Morphy Richards, Inc.)

meets a weak solution of ammonia flowing down through same. Due to the ammonia's great affinity for water, practically all of the ammonia vapor is absorbed and carried down to the reservoir, and the relatively pure hydrogen rises through the absorber and re-enters the gas heat exchanger, so completing the cycle.

"The strong ammonia liquid from the absorber passes down through the reservoir to the bottom of the boiler by way of the liquid heat exchanger where it is preheated and then passes to a small bore tube, the pump tube. The heat supplied to the boiler causes bubbles of ammonia gas to be formed, resulting in drops of liquid being forced out of the top of the tube into the outer boiler compartment (thermal-lift pump).

"Ammonia is again driven off into the condenser, and the weak solution sinks to the bottom of the boiler, into the liquid heat exchanger where it is cooled, and then passes up through the return tube into the absorber, which is arranged at a lower level than the pump to allow gravity feed to take place."

Importance of "Leveling"

The parts of the unit have been located so that a certain specified angle (from horizontal) will be obtained in the various tubes. If the refrigerator itself is not level, the vapor or liquid cannot flow effectively, and it efficiency will be poor; indeed; it may not freeze at all.

These gas-sealed units have the reputation of being exceedingly well built. After many years, however, they sometimes seem to have discharged, although it is likely that the chemicals separated. Or, possibly, an impurity caused a reaction so that the chemical balance became upset. In any event, that condition can be remedied by turning the unit upside down. How's that again? Rotate it 180 degrees so that the top rests on the floor, and just leave it alone for a few days. Many units have been successfully restored to operation this way.

Testing the Cold Control

Here, as with the mechanical refrigerator, you can suspect the cold control in case of incorrect cabinet temperature. Let us look at an electric-coil type of absorption unit. The input would be 115 volts ac if it plugs into a wall receptacle, or it could be 115 or less dc volts. It does not matter which or what voltage as long as you have the right bulb for your two-wire tester. The thermostat would make-or-break the feed to the resistance heater, and you can use the two-wire tester to determine if the heater is being energized simply by placing the two probes on the two terminals. If it is energized, and it does not give off any heat (which should be obvious if you feel it), the resistance wire must be broken. If it is not energized, go to the cold control and touch the terminals on it. If the bulb still does not light, there must be a broken wire some place. When there is a broken wire inside the insulation, the only way to isolate it is to check out the circuit. To do that (assuming the bulb socket has short leads) you must put an extension of, say, two feet on one of the leads. This can be done with another piece of wire with a clip on each end. Clip the loose end to one of the input terminals, and touch the probe to various points around the circuit. When you get to the point between where the bulb lights and where it does not, that is where the trouble is.

If you think that the switch contacts are not closing when they should, put the probes back on the resistance heater, and turn the cold control dial a little. If turning it, say, a quarter of an inch will energize the heater, the control is simply calibrated wrongly and would serve well enough on a slightly colder setting. If you have to give it, say, half a turn beyond the normal setting, it would appear to have the classic symptoms of a discharged bellows. By the same token, overcooling might be the result of a faulty switch.

Checking the switch cycle (i.e., the percentage of time that it runs) might appear to pose a problem because the unit would be silent in operation; but there is a way to do it. If the voltage is 115, you can use an electric clock! The clock wires have to be connected so that it only runs when the heater is energized, of course. If there is any question in your mind, just put the tester probes on the heater, turn the dial back and forth and notice if the clock runs when the bulb lights and is stopped when the bulb is out.

Heat Input

Absorption refrigerators run on a rather long cabinet-temperature cycle. That is, it takes quite a while after the heat goes on before the freezer temperature changes radically. This type of unit is not like the mechanical type in which, after the compressor starts, you can touch the evaporator and (with certain models) feel the metal cool almost within seconds. An absorption unit is designed to operate with a specified heat input. The heaters come in various wattage ratings for the different cabinet sizes and voltages encountered. If the unit was supposed to run say, 40% of the time, and, because the heater is of too small a rating, it ran 75% of the time, it follows that the freezer temperature would drop slower than is normal and the cabinet temperature would rise higher than is normal as a consequence. In an extreme case, this situation could result in a tendency to defrost, which might be mistaken as a symptom of a partly discharged switch bellows.

Although some variation in cabinet temperature and upsetting of the convection pattern might result, attention to correct heat input is mostly a matter of economy of operation. The right amount of heat is needed to have the chemicals flow in the tubes in an orderly fashion. If the heater is supposed to run on 115 volts and only 100 volts is being delivered, there would be a difference in the monthly power consumption. This is a case in which attention to proper wiring will pay dividends.

The proper wattage is supposed to be indicated on a small plate attached to the unit. The actual heat output in watts can be computed. All that is required is a voltmeter and ammeter. With the unit wired so that the current in one of the leads will flow through the ammeter, the reading in amperes multiplied by the known voltage will give an answer in watts. If it is more than 10% off, the unit needs attention.

Excessive operation with too cold a cabinet could be due to electrical leakage (short) between the cold-control leads, or the capillary-tube bulb might be loose and the heat transfer reduced by corrosion or ice. The bulb is supposed to be tight against the evaporator. The bulb would not be able to sense that the freezer was cold enough for the unit to shut off and could not indicate this condition because of poor contact.

Gas Burners

Most absorption units run on some form of domestic gas. There are four broad classifications: manufactured (city) gas, natural gas, propane, and butane. Various burners are capable of using each class. The burners are somewhat similar in appearance, but you can not substitute one for another because the opening that controls the gas flow is different in each. This "orifice" is in a small brass screw-in fitting, often called a "spud." In each class of burner, minor variations (mixtures) in the gas supply can be compensated for by changing spuds. The amount of air required for combustion differs with each class, and the manner of mixing the air with the gas before it burns (primary entrainment) varies also. Thus, you will find burners with a different internal arrangement with respect to the ports (air passages). People have made unauthorized substitutions, and, even though the burner could be adjusted so the flame looked all right, it produced carbon that filled the flues and cut down the draft, which resulted in even poorer combustion. With the wrong burner, some carbon monoxide would probably be produced (which you would not smell) and something else, which, impolitely stated, stinks to high heaven. It is a biting odor that, once smelled, will never be forgotten.

Gas-refrigerator controls are about the same as those on a gas water heater. There are some variations according to make and model, but, in general, to start, you push a button to allow a small amount of gas to flow from the "pilot" tube. This gas flow is ignited with a match. The flame will strike a "heat conductor" that will carry the heat to a bimetallic disc that, when heated, will warp, pulling its attached plunger away from a valve seat and opening the main gas supply to the burner. With the main valve open, and the thermostat not calling for operation of the unit, the gas will flow through a bypass and support a small flame (the minimum flame) in the burner. The flame at the "pilot" tube will ignite the burner, and, at this point, the button can be released. The burner flame will maintain enough heat on the heat conductor so that it will keep the valve open.

When the thermostat bellows expands, it has the effect of opening the gas port for a larger, or maximum, flame. The two flame sizes (usually called low and high) are regulated by screws. How much flame you will have with a particular screw setting is relative to the pressure of the gas in the supply line. For efficiency and economy of operation, the flame sizes are supposed to be held within very close limits.

The Manometer

The flame size is not a matter of dimensions, it is what you get with any given pressure. To make pressure adjustments, you need a manometer like the one shown in Fig. 9-2. This manometer is just a U-shaped length of glass tubing mounted on a board that is marked in inches so that you can read the difference between the tops of the two water columns. One glass-tube end has a rubber tube on it and the other is open to the air. The rubber tube connects to an outlet on the burner or controls. If none is readily found, it is because the manufacturer has supplied a screw-in fitting for this purpose.

Fig. 9-2 Manometer or water gauge connected to thermostat (*Courtesy*, Morphy Richards, Inc.)

To take a reading, put water in the tube until it reaches the zero mark (or thereabout). With the gas outlet open, the gas will flow through the tube and push one column down and the other up. If one reads 3 inches above zero and the other 1½ inches below zero, you have 4½ inches of gas pressure. The pressure in the main intake pipe is supposed to be 11 inches for propane, for example, and should hold very close to that as the high flame goes on and off. Readings for other gases vary so much that it is not even worthwhile to give the general order of magnitude. You will have to get the correct reading from the dealer or manufacturer.

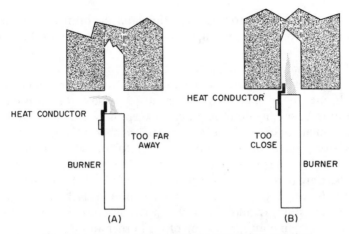

Fig. 9-3 Burner placement for vertical flue opening

The main service problems on absorption units derive from an accumulation of dirt (carbon) or the burner going out. The two are somewhat related. If the heat conductor gets too much carbon on it, the carbon will cut down the heat transfer and allow the conductor to cool off, thereby closing the bimetallic disc and shutting off the gas supply.

In Fig. 9-3A, we see a burner placed too low, with the result that the flue is not hot enough, there is little draft, and the heat dissipates to the sides. In Fig. 9-3B, on the other hand, we see what happens if the burner is too high; the burner restricts the movement of the air, which the flame needs, and the flame is "dirty." Smoke is carried up into the flues, which is the passage in and around the sealed unit, and the flues become covered with carbon, which further reduces the draft and causes still poorer combustion. This situation calls for a chimney cleaning. Using a long-handled brush (the kind used on the old coal-burning furnaces works fine) or a cord with a rag secured in the middle, you should clean the flues from top to bottom. (It will be necessary to remove the burner.)

Other problems are encountered with a horizontal flue opening. In Fig. 9-4A, we see a situation in which the heat does not enter the flue. In

Fig. 9-4 Burner placement for horizontal flue opening

Fig. 9-4B, the burner is too near the flue, and the draft carries the flame away from the conductor, which, of course, could result in the burner going out.

Then again, it might happen that the heat conductor was adjusted during the summer and much lower temperature of the air drawn from the floor during the winter was sufficient to reduce the heat-conductor temperature below the point required to keep the valve open. In this case, the heat conductor can be bent for best performance. Of course, it is not impossible for a speck of rust to be carried into the valve and lodge in the bypass opening. This situation can be remedied by cleaning.

There are two reasons why the flame might be dirty and smoky, or, in other words, why it produces carbon. One reason is lack of enough air for proper combustion. In addition to the situation referred to previously, this can result from an improper setting of the burner air intake, which governs how much air mixes with the gas before it burns. In Fig. 9-5 we see that the air shutter barrel screws onto a threaded fitting. With burners of the type shown, the amount of air mixed in is governed by turning the barrel. When the burner is turned all the way to the bottom, the flame (for lack of enough air) will be yellow and smoky. Rotating the barrel, so that it goes up to the top of the shaft, on the other hand, will give the flame too much air and it

Fig. 9-5   Gas burner component assembly

will be noisy and "fluttery." The correct setting is about half-way between these two extremes. The other reason for a carbonizing condition is that the flame is hitting the side of the flue. To correct this, the burner can be moved to center the flame in the flue opening.

Absorption units are being used in industry to utilize heat that would be otherwise wasted. Because of the increasing costs of domestic fuel gases, however, these units are losing favor for use in home refrigeration.

Many recreational vehicles make use of an absorption refrigerator, which usually operates either on 12-volt current or propane gas.

Glossary

Glossary

Bulb: The end portion of the small tube on a thermostatic switch. In the early days the tube was enlarged at the end; most modern switches have it coiled or merely looped.

Capillary (tube): A tube of very small diameter. The term is applied to the tube on a thermostatic switch, the spiral passage in a restrictor, the tubes from the restrictor to the evaporator (where used), and may also refer to the restrictor and liquid line together.

Carbon: An element, black and sooty, formed by smoke from a dirty flame.

Carbonized oil: A lubricant that has become sticky.

Charge: The specified amount of refrigerant in a system. Putting in the refrigerant is called "charging," and when it has leaked out or is removed, the system is "discharged."

Circuit: The path taken by a drop of Freon as it makes its way through the various parts of the system and returns to the point from which it started. An electrical circuit is a path through which a current is flowing.

Coil: A length of wire or tubing that is wound in a circular fashion like thread on a spool. In an air-cooled condenser or evaporator, the coil is the tubing, which is arranged in a number of loops.

Cold control: Same as thermostatic switch.

Components: In a refrigerator, the sealed unit, cold control, and relay. In a sealed unit, the motor-compressor, evaporator, and condenser.

Compressing: Causing gas or air to occupy a smaller space by exertion of pressure.

Condensate: The water that drips from an air-conditioner's cooling coil.

Condensing: Causing a vapor to become a liquid by application of pressure or lowering its temperature.

Cycle: The period of time required for a unit to start, run, shut off, stand idle, and start again.

Energize: To apply power to something electrical, such as a solenoid coil. If it has no current running through it because it is shut off, or because of a broken wire, etc., it is said to be de-energized.

Evacuate: To use a vacuum pump to remove the Freon and air, if any, from the interior of a refrigerating unit.

Evaporator: The part of the system in which the liquid Freon "boils" and turns to a vapor.

Fins: Pieces of sheet metal secured to tubing to help it give off heat, or to cool the air, as the case may be.

Freon: The most common refrigerant.

Gas: A general term applied to refrigerants. In an absorption-type refrigerator, it would refer to the fuel that is burned to produce heat that produces cold.

High side: The part of the system where a high pressure prevails, i.e., in the compressor, condenser, liquid line, or restrictor.

Liquid line: The tube that carries liquid Freon from the condenser to the restrictor. It is much smaller than the suction tube that carries the Freon vapor back.

Low side: The part of the system where low pressure prevails, i.e., in the evaporator and the suction line leading back to the compressor.

Pass: One loop of tubing.

Pressure, back: The pressure found on the low side.

Pressure, head: The pressure found on the high side.

Refrigerant: The fluid (liquid or vapor) that is used to create cold by absorbing heat at a low temperature and pressure in one place and giving off the heat at a higher temperature and pressure at another place.

Relay: A device to help the motor to start.

Resistance heater: An electric heater that has one or more wires that become red hot when energized because of the resistance to the flow of current by the wires.

Restrictor: A device that keeps the liquid Freon from running into the evaporator too fast by making it go through a tube of small diameter.

Sublimation: Process by which a solid is transformed to the gaseous state without passing through the liquid state.

Suction line: See Liquid line.

System: The interior passages through which the Freon passes as it makes its way around the circuit.

Thermally actuated switch: A device that will open (or close) contacts when the temperature changes. Sometimes called a *thermo-switch.*

Unit, open: The obsolete construction in which the component parts could be removed by the use of wrenches.

Unit, sealed: The modern construction in which the motor and compressor are in a welded shell and all tubing joints are soldered.

Vapor: Any gaseous substance. Any liquid will "boil away," or vaporize, at some temperature or other. Some solids, e.g., "dry ice," will sublimate, i.e., go from a solid to a vapor without becoming a liquid in between.

Index

Index